Elementary Mathematics for Economists

Elementary Mathematics for Economists

by

Caroline Dinwiddy

Volume I in the Series of Undergraduate Teaching Works in Economics being developed at Makerere College, University of East Africa

Nairobi
OXFORD UNIVERSITY PRESS
Lusaka New York London
1967

Oxford University Press, Ely House, London W. 1
GLASGOW NEW YORK TORONTO MELBOURNE WELLINGTON
CAPE TOWN SALISBURY IBADAN NAIROBI LUSAKA ADDIS ABABA
BOMBAY CALCUTTA MADRAS KARACHI LAHORE DACCA
KUALA LUMPUR HONG KONG TOKYO

The publication costs of this series have been subsidised by a generous grant from the Rockefeller Foundation.

Acknowledgement is made to the Cambridge University Press, for their permission to reproduce the tables that appear at the back of this book from *Four Figure Tables* by C. Godfrey and A. W. Siddons.

Printed and Made in East Africa

FOREWORD

In developing new curriculum and course work in economics during 1963-65 at Makerere College, University of East Africa, the staff felt keenly the almost-universal weakness of economics students in the realm of mathematics. Few students had more than O level mathematics, having taken mostly non-mathematical Arts subjects in Senior Secondary School, and some mature-age entry students were 10-15 years away from even that. A rudimentary knowledge of basic mathematics is absolutely essential even for first-year economics work today. Something had to be done.

One solution would have been to require economics students to take first-year mathematics as one of their three subjects. But many, if not most, could not qualify, and would not have wanted to give up an Arts or Social Science subject for mathematics if they could; furthermore, that was more mathematics than they really needed.

The alternative was to incorporate into first-year economics work some lectures on elementary mathematics so that a student would at least feel "at home" with the use of index numbers, percentage growth rates, functional relationships expressed graphically or algebraically, and even a little calculus so that the "marginal" concept in economics would have meaning.

The experiment was tried and seemed to work so well, with many students becoming enthusiastic, that we decided to offer an optional further stage in the last term, primarily for students contemplating doing further work in economics, with some study of probability theory, matrices, and difference equations useful for second-year work in economic theory and in statistics.

These lectures were given in 1964-65 by Mrs. Caroline Dinwiddy, who also set up weekly exercise papers. The programme as developed by her was so outstanding in the opinion of all of us working with her that we encouraged her—indeed begged her—to write up her notes in readable form so that we could incorporate them as a separate volume in a series of texts we were developing for African students—texts which hopefully would provide more meaningful material for the African student than those which stressed the manipulation of Bank Rate or study of the effects of automation on employment, topics stressed in the only books available in English. Mrs. Dinwiddy's volume can stand by itself, but it also forms a vital complement to other works in our series, which draw on and use the mathematics developed here.

When it was decided to go beyond mimeographing and publish some of the books in our series, Mrs. Dinwiddy's volume was bound to be included. Nowhere that I know of is the elementary mathematics necessary for the study of economics explained so lucidly and so simply. Nowhere else can the African student see how it can be applied to problems that directly concern him.

Poultney, Vermont
August 24, 1966.

Philip W. Bell

INTRODUCTION

This Outline has been written for first year economics students. It has been assumed that students have at some stage taken either the School Certificate or the G.C.E. Ordinary Level Mathematics examination, but it is fully realized that this may have been some years ago. For this reason, Part One on 'elementary algebra' has been presented for revision purposes in outline form without much written explanation. Examples have been used freely throughout and it is hoped that they are self-explanatory. Part Two, on 'introduction to calculus' has been written for students who have no previous knowledge of the subject; Part Three on 'growth' contains material useful very early on in economic study, much of which will be familiar. The last chapter in each of these first three Parts shows how the mathematics can be used in economic examples. It should be emphasized that it is assumed throughout that the student is familiar with the underlying theory—the economic examples are meant to illuminate what is known, not to teach new material. The last Part, Part Four, contains material which will be more relevant for second and third year work; here one or two economic examples are given, but fuller expositions are left for more advanced texts.

One reason for writing this book was to try to answer the question 'Why is mathematics necessary for economics students?' The immediate answer to the question is that much of the literature on economic problems today contains mathematical symbols and arguments. The mathematics involved is not usually very formidable and it is a great advantage for the student to be familiar with some of the simpler techniques. Even the most elementary economics textbooks use graphs to illustrate the basic propositions of economic theory; a little algebra can make these relationships even more revealing.

The more fundamental answer to the question of why mathematics is necessary falls into two parts. In the first place, economics is not purely a theoretical subject; it is concerned with very real problems such as how people's incomes are determined, what factors influence the prices of the commodities we buy and sell, whether a country can afford to expand its secondary school system or whether, from an economic point of view, it can afford not to. These problems cannot be dealt with in purely theoretical terms. Factual information has to be collected. How much do people earn? What have the post-war prices of coffee, tea, sisal and cotton been? What is the present enrolment in secondary schools? This material then has to be analysed: Is it reliable? Is it meaningful? Can it be used to make predictions about future trends? The process of collecting and analysing facts about the world has given rise to the branch of economics known as "econometrics" (combining "economics" with "statistics") and statistical calculations naturally require some basic mathematical knowledge.

The second use of mathematics is rather different. The science of economics collects its material from the real world, but the object of economic study is to try to distinguish basic relationships, to try to build up a body of theory which can help us understand economic events. The special role of mathematics in developing economic theory lies in the clarity and conciseness of its language. For example, the assertion, basic to classical price theory, that the quantity of a good consumers are willing to buy depends upon its price, may be expressed by the functional notation

$$Q = f(P);$$

quantity demanded is a function of price. At once we can see the exact nature of the proposition being made. The economist is assuming that the price of a good is the only determinant of the quantity demanded. All other considerations—possible changes in consumers' income, changes in tastes, prices of competing goods, etc. are explicitly ruled out—they are assumed to remain constant as quantity and price change. The limitations of the assumption are clearly seen. Suppose we wish to increase the complexity of the statement and say that the quantity of a good demanded is determined both by its price and by consumers' income: then

$$Q = f(P,Y).$$

The mathematical formulation again makes the nature of the proposition quite clear. In this simple case mathematics provides a clear and useful shorthand impression for an economic proposition.

In some of the examples to be found in the chapters on 'economic applications' mathematics will simply illuminate propositions which are already well known to the student; in others, problems are posed that could scarcely be solved without the use of simple mathematical techniques.

There is one warning that should be borne in mind. We have seen that the virtue of mathematics is that it can simplify and clarify economic propositions. This simplification has its dangers. Frequently the mathematical argument is so neat and convincing that one is tempted to forget the assumptions which were specifically excluded from the problem. We may "prove" that wages in an industry are determined by the supply of labour and the value of the output of the last man to be employed; this is true given the highly restrictive assumptions of a perfectly competitive labour market. Trade Union activities, Government legislation, the desire of the industry to establish good labour relations, have all been ignored. When assessing the conclusion of a mathematical argument it is therefore always essential to ask oneself what 'hidden assumptions' lay behind the formulation of the problem.

In writing this Outline I have been deeply indebted to Professor Bell who has encouraged and advised me throughout as well as contributing a number of the economic examples. Dr. C. Frank and Mr. E. Rado have also made valuable criticisms and suggestions. For the typing of the manuscript I am extremely grateful for the skill and perseverence of Mrs. Audrey Alexander and Mrs. Elizabeth Eddie.

I have been greatly assisted by members of the Makerere Economics Department, and I have also drawn heavily upon the following books:

R. G. D. Allen, *Mathematical Analysis for Economists* (Macmillan & Co. Ltd., London, 1938);

J. Parry Lewis, *Mathematics for Students of Economics* (Macmillan & Co. Ltd., London, 1959);

M. J. Brennan, *Preface to Econometrics* (South-western Publishing Company, Ohio, 1960);

C. J. Tranter, *Advanced Level Pure Mathematics* (English Universities Press Ltd., London, 1953).

For the more specialized topics in Part Four I have been indebted to

F. Ayres, *Matrices* (Schaum Publishing Co., New York, 1962), and to

W. J. Baumol, *Economic Dynamics* (The Macmillan Company, New York, 1951.)

The chief object of this Outline has been to make some useful mathematical concepts comprehensible to people who do not take naturally to the subject. For this reason, rigour and completeness have sometimes been sacrificed to simplicity. Those students who become interested in the mathematical approach to economics are urged to pursue the subject in some of the books listed above. Meanwhile, it is hoped that this will provide a useful introduction for undergraduates.

Makerere University College — Caroline L. Dinwiddy
1 April 1966

CONTENTS

PART FOUR. SOME EXTENSIONS OF ELEMENTARY MATHEMATICS

PART ONE
ELEMENTARY ALGEBRA

I. BASIC PROCESSES IN ALGEBRA

II. SIMPLE FRACTIONS AND FACTORS

III. SIMPLE EQUATIONS AND INEQUALITIES

IV. SIMULTANEOUS EQUATIONS AND GRAPHS

V. ECONOMIC APPLICATIONS EQUATIONS AND EQUILIBRIUM

1. BASIC PROCESSES IN ALGEBRA

1. Algebraic Vocabulary

(i) $\underline{5x}$

In the above algebraic expression:

5 is a *constant*
x is a *variable*
5 is the *coefficient* of x
(and x is the coefficient of 5)
5 and x are the two *factors* in this multiplication.

(ii) a, 3a, -2a, and 7a are known in algebra as *like terms.*
a, 2b, $3a^2$, and $-b^2$ are *unlike terms.*

(iii) $\underline{\sqrt{x}}$

The above expression represents the *square root* of x, and may be defined as that number which when multiplied by itself is equal to x.
Similarly, $\sqrt[3]{x}$, the cube root of x, is that number which when multiplied by itself three times is equal to x.

Examples
$\sqrt{16} = +4$ or -4 (See I.4).

$\sqrt[3]{27} = 3$.

(iv) $\underline{a^4}$

In the above expression

a is the *base*
4 is the *exponent*
a^4 means (a). (a). (a). (a) $= a^4$
(and should *never* be confused with 4a).

(v) *Additional Note on Exponents*

(a) $x^7 \div x^7 = x^0$ (See I.2),
but a number divided by itself is always equal to 1, therefore, by definition,
$x^0 = 1$.

(b) Negative exponents
$$\frac{1}{x} = x^{-1}$$

$$\frac{1}{x^2} = x^{-2}$$

$$\frac{1}{x^3} = x^{-3}$$

and in general

$$\frac{1}{x^n} = x^{-n} \ .$$

(c) Fractional exponents

$$\sqrt{x} = x^{\frac{1}{2}}$$

$$\sqrt[3]{x} = x^{\frac{1}{3}}$$

$$\sqrt{x^3} = x^{\frac{3}{2}}$$

$$\sqrt[3]{x^2} = x^{\frac{2}{3}}$$

and in general

$$\sqrt[n]{x} = x^{\frac{1}{n}} \ .$$

$$\sqrt[n]{x^p} = x^{\frac{p}{n}} \ .$$

2. Multiplication and Division of Numbers with Exponents

Rule. *To multiply* numbers with the same base, add the exponents.
To divide numbers with the same base, subtract the exponents.

Examples

(a) $a^3 \, . \, a^5 = a^8$

(b) $\frac{a^6}{a^2} = a^4$

(c) $a^2b(a^3+4ab) = a^5b + 4a^3b^2$

(d) $a \, . \, \sqrt{a} = a \, . \, a^{\frac{1}{2}} = a^{\frac{3}{2}} = \sqrt{a^3}$

3. Addition and Subtraction of Algebraic Terms

Only like terms can be combined by addition or subtraction.

Examples

(a) $a + 2a + 3a - 4a = 2a$

(b) $x^2 + 2xy - y + xy - 3y = x^2 + 3xy - 4y$

4. Positive and Negative Numbers

(i) *Addition and Subtraction*

To add two numbers with the *same* sign, combine their absolute values and give the answer the common sign, e.g.

$$4 + 3 = 7;$$
$$(-4) + (-3) = -7.$$

To add two numbers with *opposite* signs, subtract the smaller absolute value from the larger and give the answer the sign of the larger, e.g.

$$(+5) + (-3) = +2.$$
$$(-7) + (+4) = -3.$$

To subtract one number from another, change the sign of the subtrahend (the number to be subtracted) and proceed as in addition, e.g.

$$(+5) - (+2) = +3$$
$$(-7) - (-3) = -4$$
$$(-5) - (+3) = -8.$$

Examples

(a) Add
$$\begin{array}{r} y^2 - y + 1 \\ 3y^2 + 2y - 3 \\ \hline 4y^2 + y - 2. \end{array}$$

(b) Subtract
$$\begin{array}{r} x^2 - 2x + 1 \\ 2x^2 - 3x + 4 \\ \hline -x^2 + x - 3. \end{array}$$

(ii) *Multiplication and Division*

If the two numbers to be multiplied or divided have the same sign the answer is positive, if the two numbers have opposite signs the answer is negative.

Examples

(a) $(-x^2y)\ (xy) = -x^3y^2$.

(b) $(-ab)\ (-a^2b) = a^3b^2$.

(c) $(-x^5y^3) \div (x^2y^2) = -x^3y$.

(d) $(a^5 - ab^2) \div a = a^4 - b^2$.

5. Multiplication and Division by Polynomials

The method used is identical to that used in long multiplication or long division in arithmetic.

Examples

(a)
$$\begin{array}{r} x^2 - 2xy + y^2 \\ x - y \\ \hline x^3 - 2x^2y + xy^2 \qquad\qquad \\ - x^2y + 2xy^2 - y^3 \\ \hline x^3 - 3x^2y + 3xy^2 - y^3. \end{array}$$

(b)
$$(x-y)\ \overline{\big)\, x^3+x^2y-xy^2-y^3\ \big(}\ \underline{x^2+2xy+y^2}$$
$$\begin{array}{l} \underline{x^3-x^2y} \\ \qquad 2x^2y-\ xy^2 \\ \qquad \underline{2x^2y-2xy^2} \\ \qquad\qquad\quad xy^2-y^3 \\ \qquad\qquad\quad \underline{xy^2-y^3.} \end{array}$$

EXERCISE 1.

A. Simplify by combining like terms:
1. $3a + 4b - 5a + 4c - 6b$.
2. $x - y + z + 4x - 4y - 3z$.
3. $3m - 4n + 4p - 6m - 8n - 4p$.
4. $2abc - 3b + 2c - 5a - abc - 2a - 2c$.

B. Add :
1. $2x - 3y + 4z$, $5x - y - z$, and $- x - y + z$.
2. $3x^2 - 2x + 1$, $4x^2 - x - 9$, and $6x^2 + 7x + 1$.
3. $4a + 3b - c$, $5a - c + b$, and $4b - 6a$.
4. $2x - 1$, $- 3x + 2$, $4x - 5$, and $10x - 11$.
5. $3ab - 4bc + 5cd$, $- bc + ab$, and $8ab - cd$.

C. Subtraction:
1. From $5x - 3y + 2z$, take $x + y + z$.
2. From $7a - 4b + 6c$, take $6a - 7b - 3c$.
3. Take $2a + 3b - 7c$ from $- 3a + 9b - 6c$.
4. Take $- 3x^3 - 2x^2 + 7$ from $5x^2 - 2x - 9$,
5. Subtract $ab - ac + 5$ from 0.

D. Multiply:
1. $x^2 (x^2y)$.
2. $- xy (x^2y^2)$.
3. $- a (a^3 + a^2 - a)$.
4. $(a - b) (a^2 - 2ab + b^2)$.
5. $(2x + y) (3x - y)$.

E. Divide:

1. $4m^3 - 8m$ by $4m$.
2. $c^2 - c + 1$ by -1.
3. $6x^3 - 12x^2 - 30x$ by $-3x$.
4. $8x^3y^2 - 16x^2y^4$ by $4x^2y^2$.
5. $(a^2 - 13a + 12)$ by $(a - 1)$.

F. If $a = 3$, $b = -2$, and $c = 0$, find the values of:

1. $a + b + c$.
2. $a - b - c$.
3. $a^2 + b^2 + c^2$.
4. $a^3 + 4b^3$.
5. $b - a^2 - 6$.
6. $\dfrac{ab+c^2 - 6}{b^2}$

II SIMPLE FRACTIONS AND FACTORS

A. SIMPLE MULTIPLICATION AND FACTORS

1. Revision of Squares and Square Roots.

Examples

(a) $(5\ s^2\ t^4)^2 = 25\ s^4\ t^8$.

(b) $(-a\ b^2\ c^3)^2 = a^2\ b^4\ c^6$.

(c) $\sqrt{9\ m^2\ x^6} =$ *either* $3\ m\ x^3$

or $-3\ m\ x^3$.

2. Mental Multiplication by Monomials (Expressions with One Term) and Binomials (Expression with Two Terms).

Examples

(a) $(3ab)\ (5a^2 + 4ab - b^2)$.

$= 15\ a^3b + 12\ a^2b^2 - 3\ ab^3$.

(b) $(3x - 1)\ (2x + 3)$

$= 6x^2 + 7x - 3$.

3. Factoring Polynomials (Expressions with Many Terms).

(i) A common monomial factor may be removed from a polynomial.

Example

$16x^3y^4 + 8x^2y^2 - 4x^2y$

$= 4x^2y\ (4xy^3 + 2y - 1)$.

(ii) A trinomial (expression with three terms) with only one variable may be factored into two binomials.

Examples

(a) $h^2 + 7h + 12 = (h + 3)\ (h + 4)$

(b) $c^2 + c - 20 = (c + 5)\ (c - 4)$.

(c) $y^2 - 6y + 8 = (y - 2)\ (y - 4)$.

(d) $8x^2 - 22x + 15 = (2x - 3)\ (4x - 5)$.

B. ALGEBRAIC FRACTIONS

The rules governing the use of algebraic fractions are identical to those used for ordinary fractions.

1. Simplification of Algebraic Fractions

Fractions may be simplified by removing a common factor from numerator and denominator.

Examples

(a) $\frac{14ax}{21a^2} = \frac{2x}{3a}$ (7a is a common factor).

(b) $$\frac{a^2 - 2a - 24}{5a - 30} = \frac{(a - 6)(a + 4)}{5(a - 6)} = \frac{a + 4}{5}.$$

(a − 6 is a common factor).

2. Addition and Subtraction of Algebraic Fractions

Fractions have to have a common denominator before they can be added or subtracted.

Examples

(a) $\frac{x + y}{2} + \frac{x - y}{3}$ (Lowest Common Denominator is 6)

$$= \frac{3(x + y)}{6} + \frac{2(x - y)}{6}$$

$$= \frac{5x + y}{6}.$$

(b) $\frac{3a - 2b}{ab} - \frac{3b - a}{b^2}$ (L.C.D. $= ab^2$)

$$= \frac{b(3a - 2b)}{ab^2} - \frac{a(3b - a)}{ab^2}$$

$$= \frac{a^2 - 2b^2}{ab^2}.$$

3. Multiplication and Division of Fractions

Examples

(a) $\frac{x^2 - 1}{x^2 - 2x} \cdot \frac{3x - 6}{4x + 4}$

$$= \frac{(x + 1)(x - 1)}{x(x - 2)} \cdot \frac{3(x - 2)}{4(x + 1)}$$

$$= \frac{3(x + 1)(x - 1)(x - 2)}{4x(x - 2)(x + 1)}$$

$$= \frac{3(x - 1)}{4x}.$$

(b) $\frac{ab}{a^2 - b^2} \div \frac{1}{a - b}$

$$= \frac{ab}{(a - b)(a + b)} \cdot \frac{a - b}{1}$$

$$= \frac{ab}{a + b}.$$

4. Simplification of Complex Fractions

Example

$$\frac{\frac{a^2 - b^2}{b}}{\frac{a + b}{3}}$$

$$= \frac{a^2 - b^2}{b} \div \frac{a + b}{3}$$

$$= \frac{(a - b)(a + b)}{b} \cdot \frac{3}{a + b}$$

$$= \frac{3(a - b)}{b}$$

5. The Three Signs of a Fraction

Every fraction has three signs associated with it; the sign of the numerator, the sign of the denominator, and the sign of the fraction proper. When these signs are omitted they are understood to be plus. Thus the fraction $\frac{3}{4}$ means

$$+\frac{+3}{+4}.$$

Consider the four fractions below:

$$+\frac{+8}{+2} = +(+4) = 4 \qquad -\frac{-8}{+2} = -(-4) = 4$$

$$+\frac{-8}{-2} = +(+4) = 4 \qquad -\frac{+8}{-2} = -(-4) = 4$$

The value of each fraction is 4. Each of the four fractions may be obtained from any one of the others by changing two of the three signs.

Rule. Any two of the three signs of a fraction may be changed without changing the value of the fraction.

Example

Change $\frac{a - b}{c - d}$ to an equal fraction whose denominator is $d - c$.

$$\frac{a - b}{c - d} = \frac{b - a}{d - c}$$

$$\text{or} = -\frac{a - b}{d - c}$$

EXERCISE 2.

A. Factor the polynomials which are not prime:

1. $x^2 - 3x - 10.$
2. $p^2 - 6p + 9.$
3. $6x^3y - 15x^2y - 3xy.$
4. $c^2 - 4c - 4.$
5. $p^2 - p - 56.$
6. $r^2 - 7r - 60.$
7. $9 - 4x^2.$
8. $a^2 - 14a + 49.$
9. $2c^2 - 17c + 35.$
10. $a^2 - 64b^2.$
11. $x^2 + 8x + 12.$
12. $12a^2 - 35ab.$
13. $a^2 - 6ab^2 + 9b^4.$
14. $16x^2 + 8xy + y^2.$
15. $1 + x^2.$

B. Combine:

1. $\frac{x}{y^2} - \frac{y}{x^2}$
2. $\frac{x+1}{1-x} + \frac{x-1}{1+x}$
3. $\frac{3}{x+4} - \frac{x-3}{2x^2+7x-4}.$
4. $\frac{4}{a-2} - \frac{3a+2}{a^2-4}$

C. Multiply or divide as indicated:

1. $3a \cdot \frac{a+2}{6a^2+12a}$
2. $\frac{x^2-1}{x^2-2x} \cdot \frac{3x-6}{4x+4}$
3. $\frac{x^4-y^4}{x+y} \div \frac{x^2-y^2}{4x+4y}$
4. $\frac{ax+ay}{x-3y} \div \frac{ax^2-ay^2}{x^2-4xy+3y^2}$

D. Simplify:

1. $\frac{m + \frac{m}{n}}{\frac{1}{n} + \frac{1}{n^2}}.$
2. $\frac{a - 2 + \frac{3}{a}}{1 + \frac{1}{a}}$

III. SIMPLE EQUATIONS AND INEQUALITIES

A. EQUATIONS WITH ONE UNKNOWN

1. Nature of Equations

Equations may be of two kinds:

(i) Identical, e.g.
$$3x + x = 5x - x.$$
(ii) Conditional, e.g
$$5x + 2 = 22$$

Only conditional equations have *roots* and can be solved.

2. The Degree of an Equation

The degree of an equation is determined by the highest power (or exponent) of the unknown present in the equation.

(i) $3x + 4 = 0$ is a first degree or *linear* equation.
(ii) $x^2 + x + 12 = 0$ is a second degree or *quadratic* equation.
(iii) $x^3 - x = 0$ is a third degree or *cubic* equation.
(iv) $x^n = 0$ is an equation of the nth degree.

3. Solution of Equations: General Notes

Since all equations by definition express an equality of the left hand side with the right hand side, all operations performed upon one side must be performed upon the other if the equality is to be maintained. Any number may be added to, or subtracted from, one side of an equation, provided the same number is added to, or subtracted from, the other side. Similarly, both sides may be multiplied or divided by the same number, or be raised to the same power, or the square root of both sides may be taken. When solving equations, particular care should be taken over the removal of brackets and fractions.

4. Solution of Linear Equations

Examples

(a)
$$\begin{aligned} 3 - 5(2x + 1)(x - 3) &= 43 - 10x^2 \\ 3 - 5(2x^2 - 5x - 3) &= 43 - 10x^2 \\ 3 - 10x^2 + 25x + 15 &= 43 - 10x^2 \\ 25x &= 25 \\ x &= 1. \end{aligned}$$

(b) $\dfrac{x+3}{x-1} - \dfrac{3}{x} = 1$

(Multiply by L.C.D. x(x − 1)),

$$x(x+3) - 3(x-1) = x(x-1)$$
$$x^2 + 3x - 3x + 3 = x^2 - x$$
$$x = -3.$$

(c) $\cdot006a + \cdot296 = \cdot15a - \cdot07a$

$$6a + 296 = 150a - 70a$$
$$296 = 74a$$
$$a = 4.$$

5. Solution of Quadratic Equations

(i) *By Taking the Square Root*

Example

$$x^2 = 9$$

Answer $x = +3$ or -3.

(ii) *By Factoring*

(if $xy = 0$, then either $x = 0$, or $y = 0$)

Example

$$x^2 + x - 12 = 0$$
$$(x+4)(x-3) = 0$$

Either $x + 4 = 0$

$$x = -4$$

or $x - 3 = 0$

$$x = 3$$

Answer $x = -4$ or 3.

(iii) *By Formula*

If $ax^2 + bx + c = 0$

Then $x = \dfrac{-b \pm \sqrt{b^2 - 4ac}}{2a}$

Example

$$3x^2 - 5x = 2$$
$$3x^2 - 5x - 2 = 0$$

Therefore $a = 3$

$$b = -5$$
$$c = -2$$

$$x = \frac{5 \pm \sqrt{25 + 24}}{6}$$

$$x = \frac{5 \pm \sqrt{49}}{6}$$

$$= \frac{5 \pm 7}{6}$$

$$\text{either } x = \frac{5 + 7}{6} = 2$$

$$\text{or } x = \frac{5 - 7}{6} = -\frac{1}{3}$$

Answer $x = 2 \text{ or } -\frac{1}{3}.$

6. Solution of Literal Equations

Example

Solve for x:

$$\frac{ax}{b} - c^2 = y$$

$$\frac{ax}{b} = y + c^2$$

$$ax = b(y + c^2)$$

Answer $$x = \frac{b(y + c^2).}{a}$$

B. SIMULTANEOUS EQUATIONS*

1. Introduction

Sets of simultaneous equations can normally be solved by algebra if the number of variables is equal to the number of equations. Thus to solve for two unknowns, two equations are necessary; to solve for three unknowns, three equations are necessary; and in general, to solve for n unknowns, n equations are necessary. It may seem obvious that it is not possible to solve for n unknowns with (n — 1) equations, but to have (n + 1) equations may be equally unsatisfactory. Even when the number of equations is equal to the number of unknowns a solution may not be forthcoming. This is illustrated by the following pairs of equations. Pairs of linear equations (equations of the first degree) may be:

(i) Inconsistent, e.g.

$$x - y = 2$$
$$x - y = 5.$$

**Another method of solving sets of simultaneous equations may be found in* *Part Four, XVII.*

(ii) Dependent, e.g.

$$2x + 3y = 6$$
$$6x + 9y = 18.$$

(iii) Independent and consistent, e.g.

$$x + y = 5$$
$$x - y = 1.$$

Solutions can only be found for sets of equations which are both independent and consistent.

2. Solution of Two Independent, Consistent Linear Equations

(i) *By Elimination of One Variable by Addition or Subtraction*

Example

$$5x + 2y = -5 \qquad (1)$$
$$-3x + 4y = 29 \qquad (2)$$

Multiply the first equation by 2:

$$10x + 4y = -10 \qquad (1) \times 2$$
$$-3x + 4y = 29 \qquad (2)$$
$$13x = -39 \text{ (by subtraction)}$$
$$x = -3$$

Substitute (– 3) for x in first equation:

$$5(-3) + 2y = -5 \qquad (1)$$
$$-15 + 2y = -5$$
$$y = 5$$

Answer $x = -3, y = 5.$

(ii) *By Substitution*

Example

$$3x - 4y = 5 \qquad (1)$$
$$x + 7y = 10 \qquad (2)$$

From the second equation $x = 10 - 7y$; therefore substitute $(10 - 7y)$ for x in the first equation;

$$3(10 - 7y) - 4y = 5 \qquad (1)$$
$$30 - 21y - 4y = 5$$
$$30 - 25y = 5$$
$$y = 1$$

Substitute 1 for y in first equation:

$$3x - 4(1) = 5 \qquad (1)$$
$$3x - 4 = 5$$
$$x = 3$$

Answer $x = 3, y = 1.$

3. Solution of Linear Equations with three Unknowns

To solve a set of linear equations with three unknowns, eliminate one variable and then proceed as above.

Example

$$2a + 3b + 3c = 0 \qquad (1)$$
$$a + b + 2c = -3 \qquad (2)$$
$$3a - b - c = 11 \qquad (3)$$

Add (2) and (3):

$$a + b + 2c = -3 \qquad (2)$$
$$3a - b - c = 11 \qquad (3)$$
$$4a + c = 8 \qquad (4)$$

Multiply (2) by 3, and subtract from (1):

$$2a + 3b + 3c = 0 \qquad (1)$$
$$3a + 3b + 6c = -9 \qquad (2) \times 3$$
$$-a - 3c = 9 \qquad (5)$$

Multiply (4) by 3, and add to (5):

$$12a + 3c = 24 \qquad (4) \times 3$$
$$-a - 3c = 9 \qquad (5)$$
$$11a = 33$$
$$a = 3$$

Substitute 3 for a in (4):

$$4(3) + c = 8 \qquad (4)$$
$$12 + c = 8$$
$$c = -4$$

Substitute 3 for a, -4 for c, in (1)

$$2(3) + 3b + 3(-4) = 0 \quad (1)$$
$$6 + 3b - 12 = 0$$
$$b = 2$$

Answer $a = 3$; $b = 2$; $c = -4$.

4. Solution of Simultaneous Equations with One Linear and One Quadratic Equation

Simultaneous equations of a more complex kind may also be solved by substitution. Consider two equations, one linear and one quadratic. (Note: there will be four answers, representing two solutions).

Example

$$x^2 + 4y = 20 \qquad (1)$$
$$y - 2x = 8 \qquad (2)$$

Substitute $(8 + 2x)$ for y from the second equation in the first equation:

$$x^2 + 4(8 + 2x) = 20 \qquad (1)$$
$$x^2 + 32 + 8x = 20$$
$$x^2 + 8x + 12 = 0$$
$$(x + 6)(x + 2) = 0$$
$$\text{either} \quad x = -6$$
$$\text{or} \quad x = -2$$

If $x = -6$, the second equation becomes:

$$y - 2(-6) = 8 \qquad (2)$$
$$y + 12 = 8$$
$$y = -4$$

If $x = -2$, the second equation becomes:

$$y - 2(-2) = 8 \qquad (2)$$
$$y + 4 = 8$$
$$y = 4$$

Answer There are two solutions:

$$\text{Either } \left.\begin{array}{l} x = -6 \\ y = -4 \end{array}\right\} \quad \text{or } \left.\begin{array}{l} x = -2 \\ y = 4. \end{array}\right\}$$

C. INEQUALITIES

1. Definitions

$a > b$ means "a is greater than b".
$a < b$ means "a is smaller than b".
$a \geq b$ means "a is greater than, or equal to, b".
$a \leq b$ means "a is equal to, or smaller than, b".

2. Rules for the Use of Inequalities

(i) An inequality will still hold after each side has been increased, diminished, multiplied or divided by the same POSITIVE quantity.

Example

If $a > b$, then, provided $c > 0$,

$$(a + c) > (b + c)$$
$$(a - c) > (b - c)$$
$$ac > bc$$
$$\frac{a}{c} > \frac{b}{c}.$$

(ii) Any term in an equality may be moved from one side to the other provided that its sign is changed.

Example

If $a - c > b$

$a > b + c.$

(iii) If the sides of an inequality are interchanged then the sign must also be changed.

Example

If $a > b$

Then $b < a.$

(iv) If both sides of an inequality are multiplied or divided by the same negative quantity then the sign of the inequality must be reversed.

Example

If $a > b$, and $p < 0.$

Then $pa < pb$

and $\frac{a}{p} < \frac{b}{p}$

(v) If the two sides of an inequality, each having the same sign, are inverted (i.e. turned upside down) then the sign must be reversed.

Examples

(a) If $\frac{a}{c} > \frac{b}{d}$

Then $\frac{c}{a} < \frac{d}{b}.$

(b) If $a < b$

Then $\frac{1}{a} > \frac{1}{b}.$

EXERCISE 3.

A. Solve the following equations:

1. $8 - 3(x + 4) = 2.$
2. $(x + 1)^2 - (x - 2)^2 = 21.$
3. $x = 4{\cdot}25 - {\cdot}25(2x - 1).$
4. $3{\cdot}2y + {\cdot}4 = {\cdot}8y + 10$

B. Solve the following equations:

1. $\dfrac{c-12}{10} = \dfrac{3(2c+1)}{5} - \dfrac{(3c-6)}{2}$
2. $\dfrac{1}{x} + \dfrac{1}{x+3} = \dfrac{2}{x-3}$
3. $\dfrac{x+1}{x-1} - \dfrac{x-1}{x+1} = \dfrac{5x+2}{x^2-1}$
4. $\dfrac{x-5}{2} + \dfrac{x+5}{2} = 0$

C. Solve by factoring:

1. $c^2 - 12c - 13 = 0$
2. $x(2x+3) = 3x + 50$
3. $x^2 + x - 12 = 0$
4. $x^2 + 4x - 117 = 0$

D. Solve by the quadratic formula:

1. $5x^2 + 9x - 2 = 0$
2. $b^2 - 20b = 96$
3. $(x-3)^2 = x - 1$
4. $\dfrac{1}{x-1} + \dfrac{1}{x-2} = \dfrac{1}{x-3}$

E. Solve for y:

1. $2(a-y) = 5y - 19a$
2. $5by - a = b(b+2y)$
3. $ay^2 + by + c = 0$
4. $y^2 + 2(b-1) + b^2 = 0$
5. $y^2 - 3ay + 2a^2 = 0$

F. Solve for the unknowns in the following sets of equations:

1. $\dfrac{7a}{4} + \dfrac{5b}{3} + 3 = 0$

   $\dfrac{a}{2} - \dfrac{b}{3} = 4$

2. $12(x-y) + 11y = 14$

   $18x + 5(y+1) = 0$

3. $x + y = 8000$

   $\cdot 04x + \cdot 06y = \cdot 0525(x+y)$

4. $x + 2y + 3z = 5$

   $2x + 4y + 5z = 4$

   $3x + 5y + 6z = 3$

5. $x = y - 1$
   $x^2 - 3y = -3$

6. $2xy + 5y = -55$
   $3y - 2x = 43$

G. 1. Express the following as inequalities:

(a) The variable x has values between 3 and 7 inclusive.
(b) y is a positive number less than or equal to 10.
(c) x exceeds y by more than 2.

2. Solve each inequality for x:

(a) $3x - 8 \geq 4$

(b) $-3 < \frac{x - 5}{2} < 3$

(c) $-1 \leq \frac{3 - 2x}{5} \leq 7$

3. Prove that if $a > b$ then $\frac{a}{b} > \frac{a + x}{b + x}$
if $x > 0$, but that the reverse is true
if $x < 0$, provided that $b > 0$ and $x > -b$.

IV. FUNCTIONS, VARIABLES AND GRAPHS

1. The Concept of a Function

In an equation we are asserting a particular relationship between two or more algebraic expressions. Equations may contain any number of variables; only under certain conditions can an equation or a system of equations be solved (see III). In other cases an equation simply expresses a relationship between two or more variables. For example, the equation

$$y = x^2 - x + 12$$

is true for an indefinite number of possible combinations of x and y.

Another way of describing an equation of this kind is to say that it shows a functional relationship. In the above expression y is a *function* of x ; that is, the value of y depends upon the value taken by the variable x. When x is 2, y is 14; when x is 3, y is 18, etc. Two more functional relationships are illustrated by the following expressions:

$$y = \frac{x^3 + 2x}{3x - 8}$$

$$y = x^2 - 2ax + a^2.$$

In the first case y is a function of x; in the second, y is a function of x and a. In mathematical terminology

$$y = f(x)$$

$$\text{and } y = f(a,x).$$

2. Explicit and Implicit Functions

Any function which is of the form

$$y = f(x)$$

is known as an *explicit* function of y. The value of y is clearly seen to depend upon the value of x, and we say that while x is an *independent variable*, y is a *dependent variable*.

Sometimes a functional relationship does not distinguish clearly between dependent and independent variables. For example,

$$2x^2 + 3xy + y^2 = 0$$

is a function of the form

$$f(x,y) = 0.$$

Such a function is known as an *implicit function.*

An explicit function can sometimes, but not always, be derived from simple implicit functions, for example, the implicit function

$$2x^2 + 3y - 8 = 0$$

yields two explicit functions:

$$y = \frac{8 - 2x^2}{3}$$

$$x = \sqrt{4 - \frac{3y}{2}}$$

3. The Range of a Variable

For many functions of the form $y = f(x)$ the independent variable x can take any value—negative, zero or positive. In such cases we say that the range of the variable x is the whole set of real numbers. Examples of this type of function are

$$y = 3x$$
$$y = 2x^2 - 5$$
$$y = x^3 - 3x^2 + 3x - 1.$$

In other cases we may wish to limit the range of the independent variable. For example in the function

$$y = \sqrt{x}$$

if y is to be a real number x must have a positive value. In this case x is limited to the range of positive numbers; the range is

$$x > 0.$$

If the movement of the independent variable is to be restricted over a certain range, then the upper and lower limits of the range should be indicated. Thus if we are interested in all values of x between $x = a$, and $x = b$, the range is

$$a < x < b$$

or, more strictly speaking,

$$a \leq x \leq b,$$

where x may be equal to either of the limits or lie between them.

In algebraic functions it is assumed that the variable is *continuous* so that there are no "gaps" in the function over the range of possible values. Thus between the values $x = 2$ and $x = 3$ there will be the values $x = 2{\cdot}1, 2{\cdot}2, 2{\cdot}3, \ldots$, and between $x = 2$ and $x = 2{\cdot}1$, there will be an infinite number of further values of x, measured in hundredths, thousandths, millionths of a unit. It should be noticed that in functions which are meant to represent real-life relationships the independent variables are often *discrete* and not continuous. If the function $T = f(F)$ is meant to represent the fact that the rate of taxation depends upon the size of the taxpayer's family, F can have the values 1,2,3,4,5, ..., but there are no intervening values. F cannot be $2\frac{1}{2}$.

4. The Graphical Representation of Functions

Many simple functions can be drawn on a graph. This technique can be useful both for bringing out more clearly the characteristics of certain types of function, and, where two or more functions are being considered, for solving sets of simultaneous equations. Before illustrating some of the more common types of function we review the basic ideas of coordinate geometry.

(i) *Cartesian Coordinates*

It is possible to locate any point on a two-dimensional surface, given two directions — a vertical and a horizontal direction.

In mathematics it is customary to measure these directions from a horizontal line (the x-axis) and a vertical line (the y-axis). These two axes cross at the *origin*. All points to the right of the y-axis give positive values for x, and all points to the left of the y-axis give negative values for x. Similarly, all points above the x-axis have positive values for y, and all points below the x-axis have negative values for y.

The two directions which locate a point are known as the *coordinates* of the point. When giving the coordinates of a point the x coordinate is always given first. The points P(5,—2) and Q(—5,4) are illustrated below.

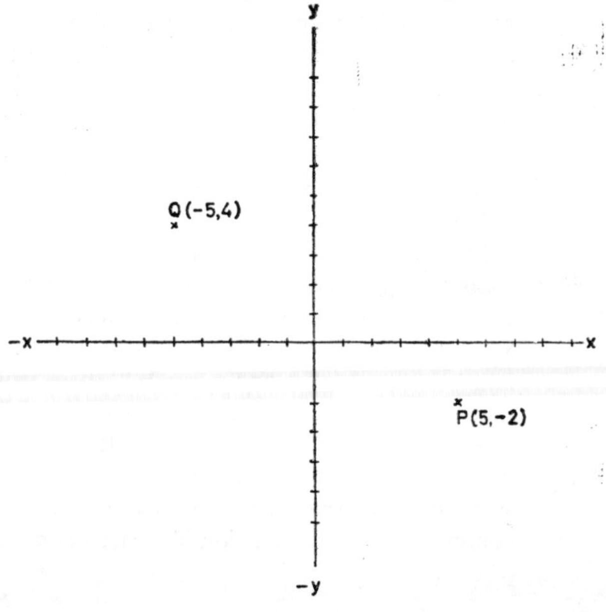

(ii) *Drawing Graphs*

Functions with two variables can be graphically represented on a two-dimensional graph. (To represent a function with three variables — for example, y = f(x,z) — a three dimensional graph with an x-axis, a y-axis and a z-axis, is necessary. This is complicated and will not be attempted in this outline). To draw the graph of any function

$$y = f(x)$$

- (i) Make a table, assigning values to x and calculating corresponding values for y.
- (ii) Draw axes; label them; mark origin 0.
- (iii) Choose a sensible and consistent scale.
- (iv) Realize that each pair of values represents the coordinates of a point lying on the curve.
- (v) Plot the points and join them.
- (vi) Label the curve.

5. Examples of Different Types of Functions

(i) *Linear functions*

The graph of a function of the first degree

$$y = ax + b$$

will always be a straight line. Consideration of this function shows that "b" represents the value of y when x = 0. Therefore (0,b) will always be one of the points on the curve. (As (0,b) is the point at which the curve crosses the y-axis, b is sometimes called the y-intercept.) "a" shows how y is related to x and determines the slope of the line. If "a" is positive y will increase as x increases, and the curve will slope upwards to the right. If "a" is negative, y will decrease as x increases and the curve will slope downwards to the right.

Examples

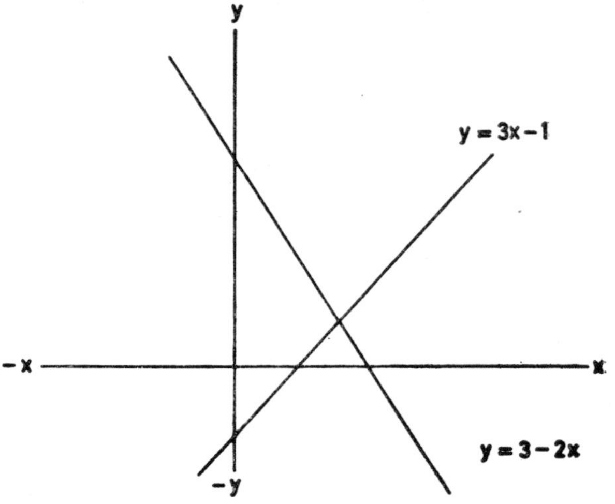

Note:

To draw the graph of a linear function it is only necessary to plot two, or preferably, three points.

(ii) *Quadratic functions*

The graph of a function of the second degree

$$y = ax^2 + bx + c$$

is U-shaped, and is known as a parabola.

Examples

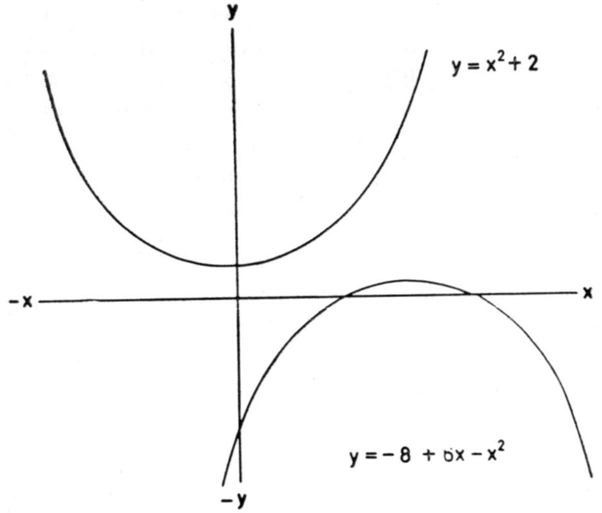

(iii) *Cubic functions*

The graph of a function of the third degree

$$y = ax^3 + bx^2 + cx + d$$

is not so easy to predict. In general, a cubic function will have either two turning points or one point of inflexion (see VI).

Examples

(a)

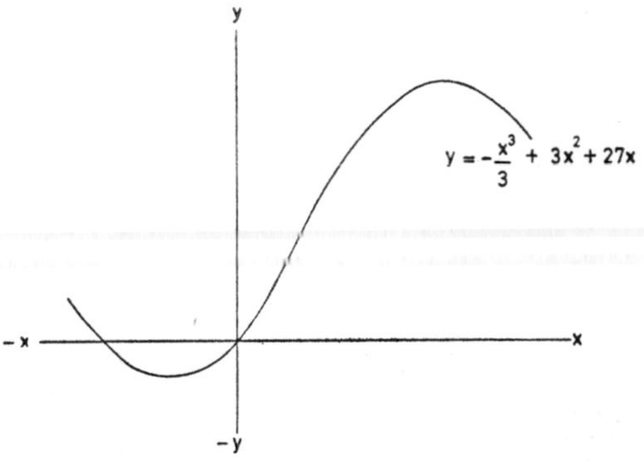

(b)

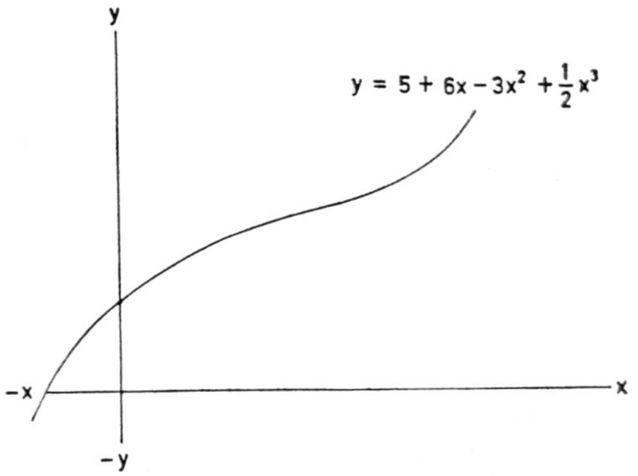

(iv) *The Rectangular Hyperbola*
The graph of a function of the form

$$y = \frac{k}{x}$$

is known as a rectangular hyperbola.

Example

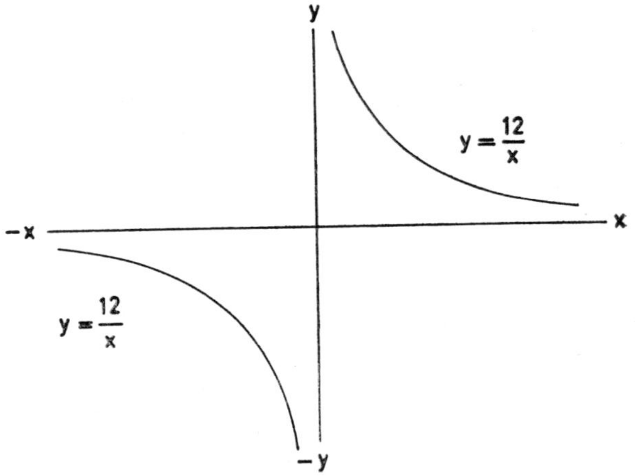

(v) *Exponential Functions*
Equations of the form

$$y = a^x$$

are known as exponential functions

Example

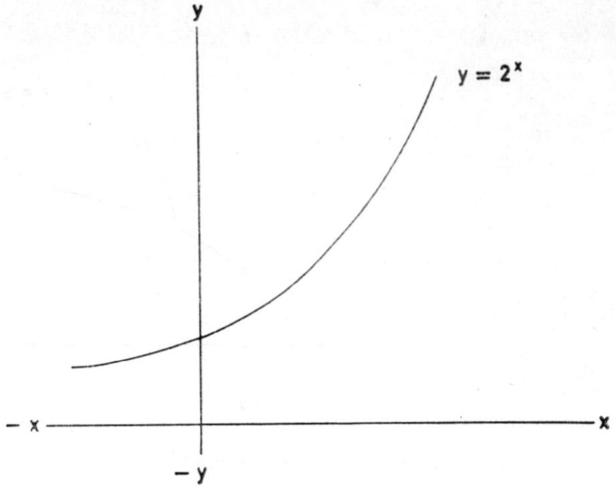

(vi) *Logarithmic Functions*
Equations of the form

$$y = \log_a x$$

are known as logarithmic functions

Example

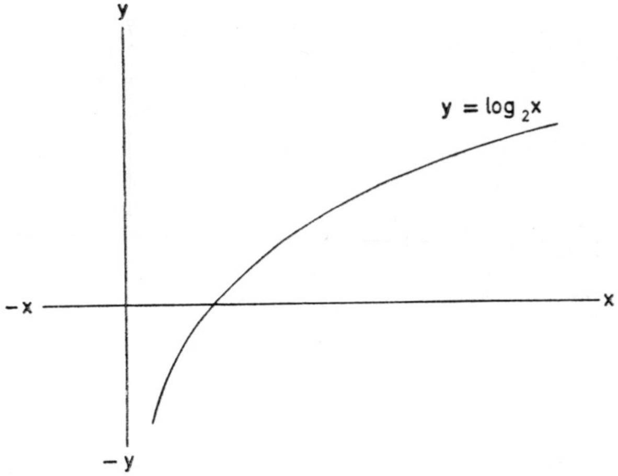

For further information about logarithmic and exponential functions, see XII.

EXERCISE 4

1. Find the explicit functions of x,y, which correspond to the following implicit functions:

(a) $3x - 4y = 0$
(b) $x - 5y + 10 = 0$
(c) $x^2 - 2y + 6 = 0$
(d) $x^3 - 4y^2 = 0$

2. Consider the following functions and answer the questions below:

$$y = x^2 + 2x$$
$$y = 3x - 1$$
$$xy = 18$$
$$y = x^3 + 4$$
$$y = 15 + 2x - x^2$$

(a) What shape do you expect the graph of the function to have?
(b) When $x = 0$, what is the value of y? What is the graphical significance of this point?
(c) When x takes large, positive values, what happens to y?
(d) When x takes large negative values, what happens to y?

3. (a) Show graphically that there is no solution to the following pair of equations:

$$y = 3x + 2$$
$$y = 3x - 1.$$

(b) Show graphically that there is no one solution to the following set of equations:

$$y = 3x + 1$$
$$y = 2x - 5$$
$$y = 7 - 2x.$$

4. On the same axes draw the graphs of the following functions for values of x from -3 to $+4$:

$$y = 2x + 6$$
$$y = x^2 - 2.$$

From the graph, read off the values of x and y which satisfy both equations.

5. Draw the graph of the function $xy = 24$ for values of x from -12 to $+12$. On the same axes draw the graph of $2y = 30 - 3x$. Give the common solutions of the two equations.

6. Draw the graph of the function

$$y = x^3 - x^2 - 6x$$

for the range

$$-3 < x < +4.$$

From the graph determine the approximate roots of the equation

$$x^3 - x^2 - 6x + 6 = 0.$$

V. ECONOMIC APPLICATIONS: EQUATIONS AND EQUILIBRIUM

A. PRICE THEORY

1. Introduction: Functions and Models

Most of the basic propositions of economic theory are concerned with functional relationships, and can be represented or formulated mathematically. In the field of price theory we may start with two very simple assumptions:

(i) The quantity of a commodity offered for sale at any given time depends upon the price. In mathematical language, the quantity supplied is a function of the price:

$$Q^s = f_1(P).$$

Furthermore, it is assumed that the quantity supplied will increase if the price rises and decrease if the price drops.

(ii) The quantity of a commodity consumers are willing to buy at any given time depends upon the price.

$$Q^d = f_2(P).$$

It is assumed that the quantity demanded will increase if the price drops and decrease if the price rises.

The problem is to find mathematical functions which would suitably describe supply and demand functions. If price is measured on the vertical axis and quantity on the horizontal axis, we know from the above information that the normal supply curve will slope upwards and the normal demand curve will slope downwards.

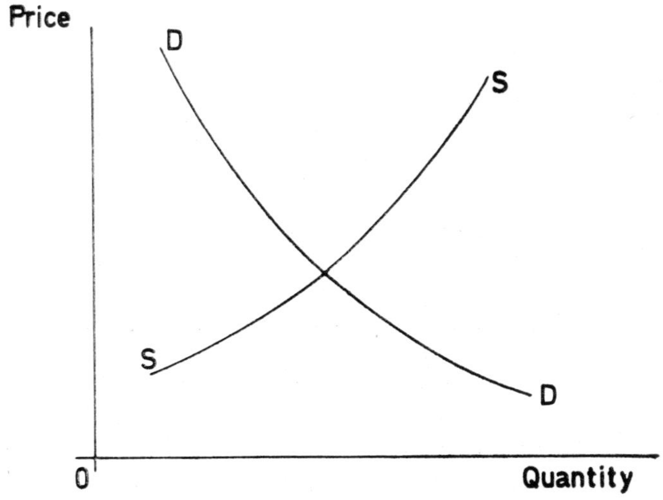

Two suitable functions would be

$$Q^s = 3P$$
$$Q^d = 40 - 2P$$

These two functions will of course produce straight line supply and demand curves. Two second degree equations which would be suitable are

$$Q^s = P^2 + 2 \quad \text{(illustrated p. 24)}$$
$$Q^d = \frac{12}{P} \quad \text{(illustrated p. 25)}$$

The first is a parabola, the second a rectangular hyperbola. In both of these functions part of the graph is irrelevant. Negative prices and quantities are not of interest to the economist—or anyone else. For this reason graphs and diagrams in economics usually show only the positive sections of the functions portrayed: the rest is simply ignored.

In the above functions P and Q are the variables of the function. The constants which determine the exact relationship of P to Q are sometimes known as the *parameters* of the functions. In the above linear demand function 40 and −2 are the two parameters, chosen we must assume from empirical observation. (See 2 below).

Having chosen two suitable functions to represent the supply and demand relationships, we proceed to set up a model. A model is simply a system of simultaneous equations describing some aspect of economic life. To find the value of the different variables contained in the model it is necessary that the number of unknowns in the model should be exactly equal to the number of equations (see III.) Let us take the second pair of supply and demand equations

$$Q^s = P^2 + 2$$
$$Q^d = \frac{12}{P}.$$

This model is not yet complete. We have two equations but three unknowns; Q^s, Q^d, and P. What we are looking for is the equilibrium situation: the values of P and Q for which the quantity offered for sale is exactly equal to the quantity purchasers wish to buy. This gives us the third equation: for equilibrium

$$Q^s = Q^d.$$

The model is now complete and by graphing the functions we can solve for P and Q and find that the equilibrium price and quantity are 2 and 6 respectively.

This type of approach which involves setting up a system of simultaneous equations and then solving to find the equilibrium values of the variables is one of the basic tools of the econometrician. Naturally, the models will normally be more complex than the preceding one, but the principles are much the same. The chief problem does not usually lie in solving the system of simultaneous equations, but rather in finding functions which accurately describe the real world, and are not just mathematicians' toys. The method of attack is demonstrated in the following example.

2. Demand Supply and Price Equilibrium: an Example

Suppose we knew (or thought we knew) that at a price of Shs. 10 the quantity demanded of a particular commodity would be 250 units, and that the quantity demanded would *increase* by 50 units for every one shilling *fall* in price below the level of Shs. 10, and *decrease* by 50 units for every one shilling *rise* in price above the level of Shs. 10.

The *demand curve* is a straight line, as shown in the diagram below. Such a linear relationship between quantity demanded (Q^d) and price (P) can be written as:

$$Q^d = a - bP.$$

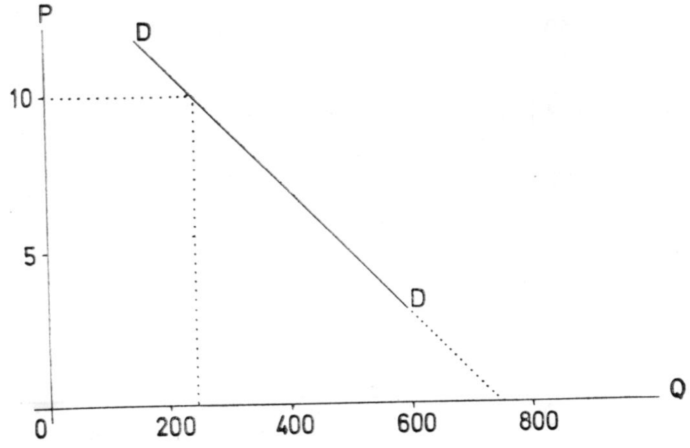

If the demand curve is extended to the x-axis (dotted extension), it would imply that some hypothetical finite quantity would be demanded if the good was a "free good". The practical realism or unrealism of such a point is of no interest to us since we are only interested in the portion of the curve represented by the solid line; but extension to the point where $P = 0$ provides us with a value for the constant "a" in the above equation, for if $P = 0$, $Q^d = a$. Furthermore, we know that $(-b)$ is the *slope* of the curve; starting at the point where $P = 0$, we know that for every one shilling rise in price, quantity demanded will fall by 50 units, i.e. $b=50$.

Even without accurately graphing the equation, then, we can write;

$$Q^d = a - 50P.$$

And since we know from the information given us that when $P = 10$, $Q^d = 250$, we can find the value of "a" by substituting these values:

$$250 = a - (50)(10), \text{ or}$$
$$a = 750.$$

So the demand equation is:

$$Q^d = 750 - 50P.$$

Now let us look at the supply side of the market. Suppose we know that if the price were as low as Shs. 5 no one would offer any of the commodity for sale, and that for every one shilling rise in price above that level, 20 (additional) units would be offered for sale on the market. The supply equation, shown in the graph below, can be written as:

$$Q^s = c + dP.$$

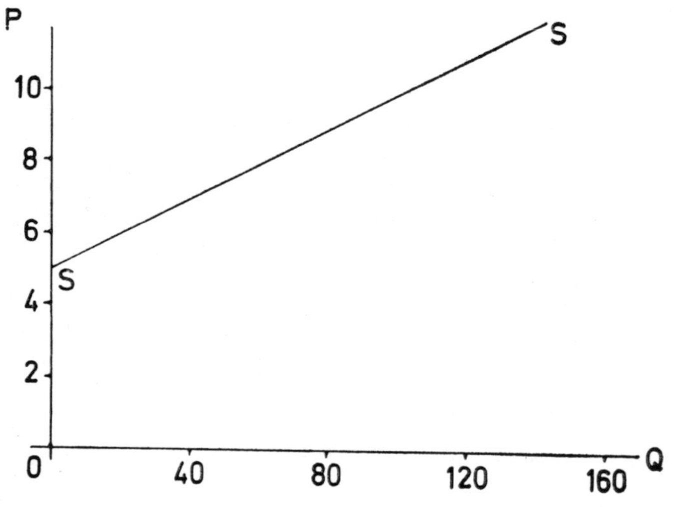

$d = 20$, i.e. the additional quantity supplied for every one shilling rise in price. Since we know that $Q^s = 0$ when $P = 5$, we can readily solve for the constant c:

$$0 = c + (20)(5), \text{ or}$$
$$c = -100$$

Thus, the supply equation is:

$$Q^s = -100 + 20P.$$

Now in equilibrium, quantity demanded must equal quantity supplied, i.e. the market must be "cleared." The *intersection point* of the two curves, shown in the third diagram, will yield the equilibrium price.

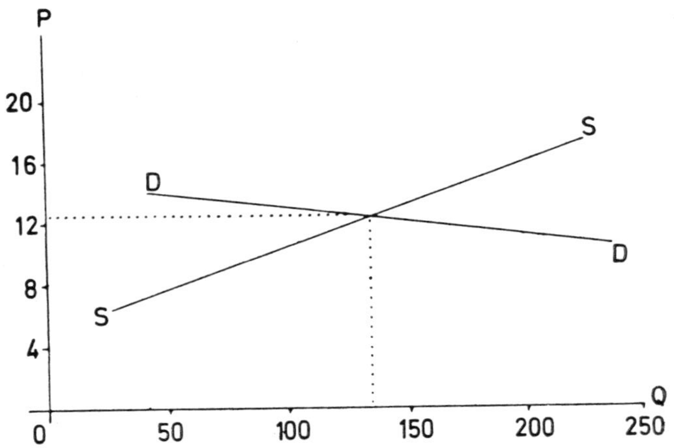

This implies that

$$
\begin{aligned}
Q^d &= Q^s, \text{ or}\\
750 - 50P &= -100 + 20P\\
-70P &= 850\\
P &= 12{\cdot}14
\end{aligned}
$$

The equilibrium price which should be established in this market is Shs. 12/14. The quantity bought and sold at that price can be determined by substituting $P = 12{\cdot}14$ into either the demand or supply equation,

$$
\begin{aligned}
Q^s &= -100 + (20)(12{\cdot}14)\\
Q^s &= -100 + 242{\cdot}80\\
Q^s\ (= Q^d) &= 142{\cdot}80.
\end{aligned}
$$

3. Endogenous and Exogenous Variables

Let us look at the linear model used in the last example. using the literal equations

$$
\begin{aligned}
Q^d &= a - bP\\
Q^s &= c + dP\\
Q^d &= Q^s.
\end{aligned}
$$

The variables in the model P and Q were interrelated and the value of one depended on the value of the other. When solving the equations we found a value for P and then found Q by substitution. Suppose that we are dealing with an agricultural product and wish to increase the realism of the model by including the monthly rainfall (R) in the supply equation so that the function becomes

$$Q^s = c + dP + eR.$$

This new variable is of a different kind from the other variables P and Q. The values of P and Q are determined within the model: they are known as *endogenous* variables. The monthly rainfall R, however, is not determined by any other variable within the system—changes in either P or Q will not affect R. Since the value of R is determined by forces outside the model it is known as an *exogenous* variable.

It is interesting to consider the graphical interpretation of a change in the exogenous variable. Let us suppose that the above supply equation

$$Q^s = c + dP + eR$$

has the following parameters:

$$Q^s = 4 + 3P + 2R.$$

In a certain month the rainfall was 2 inches, so the supply function was

$$Q^s = 4 + 3P + 2(2)$$
$$= 8 + 3P.$$

In the same month of the following year the rainfall was 3·5 inches; the supply function was

$$Q^s = 4 + 3P + 2(3 \cdot 5)$$
$$= 11 + 3P.$$

Graphing these two functions we have

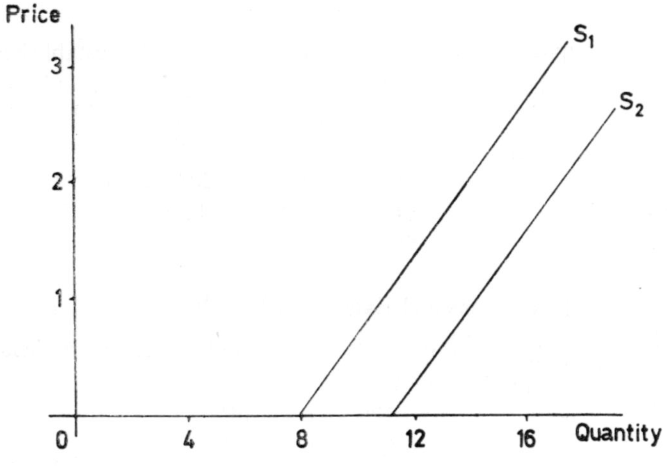

The change in the exogenous variable has led to a *shift* in the supply schedule. If, however, we take different values for the endogenous variables P and Q, then provided the parameters of the function remain unchanged, the supply curve will not shift. Different values of P and Q simply represent different points on the existing supply curve. For S_1 above

$$Q^s = 8 + 3P.$$

When P = 2, Q = 14; when P increases to 3, Q increases to 17 – another point on the same curve.

Similarly, for any demand curve of the form $Q^d = f(P)$, changes in any exogenous variables—incomes, tastes, etc.—will shift the demand curve; changes in price will not.

B. INCOME THEORY

1. The Simple Keynesian Model

National income theory may be thought of as based on two identities. The first states that for a closed economy total expenditure is the sum of expenditure on consumption goods, expenditure on investment goods, and Government spending:

$$E \equiv C + I + G.$$

The second identity describes the breakdown of the national income as it is received by households in the economy; income is either used for buying consumption goods, or saved, or taken by the Government by taxation.

$$Y \equiv C + S + T.$$

For any given period of time, total expenditure must be exactly equal to total income received:

$$E \equiv Y$$

and therefore

$$C + I + G \equiv C + S + T$$

If we assume Government spending exactly equal to the revenue from taxation so that $G = T$, then since C appears on both sides of the equation, we have the identity

$$I \equiv S.$$

These identities show that for any past period of time total expenditure must have been equal to income received and that therefore *past* Investment must have been equal to *past* Savings. However, there is no reason to assume that at the beginning of the period, the amount that businessmen *plan* to invest will necessarily be equal to the amount that householders *plan* to save. Households may plan to save more than businessmen plan to invest, or businessmen may plan to invest more than households plan to save. In the first case, deflationary pressures will lower national income until the equilibrium point is reached; in the second case, inflationary pressures will raise national income to the equilibrium level where savings are equal to investment. The economy will only be in equilibrium, with income remaining at the same level, when planned investment is equal to planned savings; in this case, total planned expenditure will be equal to total expected income.

This Keynesian argument can be illustrated very neatly by a mathematical model.

Example

If the average propensity to save in a country is given by the expression

$$S = \cdot 2Y - 50$$

and the level of investment is given by

$$I = \cdot 1Y$$

find the equilibrium level of income.

So far, the model is incomplete; we have three unknowns (S, I and Y), and only two equations. The third equation is: for equilibrium,

$$S = I$$

Solving for Y

$$\begin{aligned} \cdot 2Y - 50 &= \cdot 1Y \\ \cdot 1Y &= 50 \\ Y &= 500 \end{aligned}$$

Suppose the average propensity to save increases and the savings function becomes

$$S' = \cdot 2Y - 35.$$

If the investment function is unchanged, we should expect deflationary pressures to lower the national income. Let

$$\begin{aligned} S' &= I \\ \cdot 2Y - 35 &= \cdot 1Y \\ \cdot 1Y &= 35 \\ Y &= 350 \end{aligned}$$

These two different situations may be shown graphically:

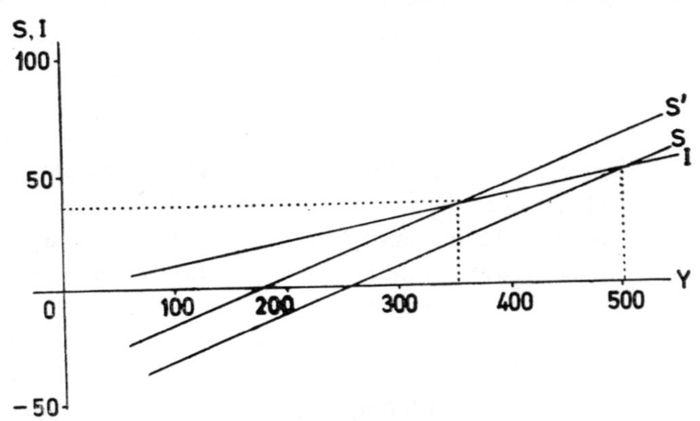

Another diagram frequently used in income theory measures national expenditure (E) on the vertical axis and income (Y) on the horizontal axis. We have seen that for equilibrium

$$E = Y$$

Points of equilibrium are to be found along the 45° line—equidistant from both axes. As national expenditure rises from an increase in consumption spending, investment spending or Government spending, the equilibrium point

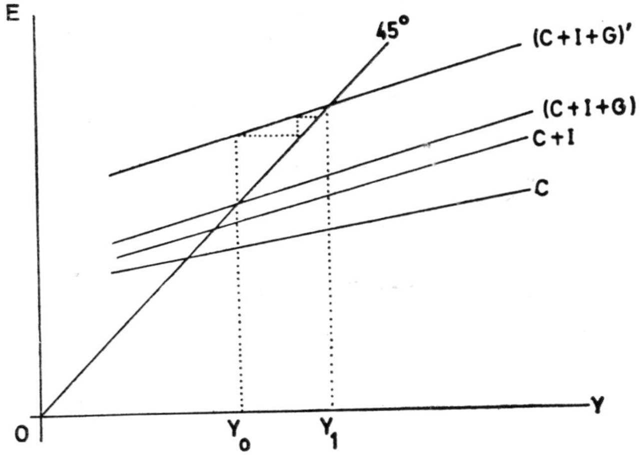

will rise and national income (measured on the horizontal axis) increases.

An important part of national income theory states—a fact that is clear from this diagram—that an increase in one of the components of the national expenditure will lead to a more than proportionate change in income (Y). This point is developed in the paragraph on the "Multiplier" in XIV.

2. A Model for Equilibrium Growth

The models constructed so far have all dealt with static situations. It is, however, possible to use a set of simultaneous equations to establish the conditions for equilibrium growth in the economy. Let us use the following symbols:

Y = national income (≡ output ≡ expenditure)
C = consumption
S = saving
c,s = the propensity* to consume and save out of income. We have, therefore,
$C = cY;\ S = sY;\ c + s = 1$
I = investment
K = the capital stock
v = the capital/output ratio (e.g. if v=4, this means that for each additional unit of output (Y) 4 additional units of capital are required. If the value of v is assumed constant so that the marginal and average capital/output ratios are the same, then for the economy as a whole

$$Y = \frac{K}{v}.$$

This implies that the stock of capital is the limiting factor

**For simplicity, no distinction is made between the marginal and average propensities to consume and save.*

on the capacity of the economy to produce output—rather than labour or natural resources.

Our knowledge of national income theory enables us to say that total expenditure

$$\begin{aligned} Y &= C + I \\ &= cY + I \end{aligned}$$

Therefore, re-arranging this equation

$$Y = \frac{I}{1 - c}$$

or

$$Y = \frac{I}{s} \qquad (1)$$

Equation (1) is the "multiplier" equation which shows how the level of investment and the propensity to save between them determine the level of income.

For output we have established that with a constant capital/output ratio

$$Y = \frac{K}{v} \qquad (2)$$

Combining these two equations we can show that expenditure out of income will lead to the full use of productive capacity if, and only if,

$$\frac{I}{s} = \frac{K}{v} \qquad (3)$$

This, then, is the concise statement of the conditions for using available resources fully. The sum total of expenditures generated $(\frac{I}{s})$ must just purchase the output produced when the capacity of the economy is fully used $(\frac{K}{v})$. If the ratio $\frac{I}{s}$ is too small so that

$$\frac{I}{s} < \frac{K}{v}$$

then demand is insufficient and the economy is working at less than full capacity. In these circumstances, "Keynesian" remedies are called for to raise output to the limit set by "capacity". The two possibilities are either to increase investment (1) or to increase consumption which means lowering the propensity to save (s). Both of these steps would tend to increase the ratio $\frac{I}{s}$ and in time "full capacity" income and expenditure would again be generated.

Unfortunately, the problem before underdeveloped countries is often a more serious one. The difficulty is rarely one of working "below capacity", but rather that even at full capacity, output per head is too low. The simple remedy of saving less (and spending more) will not suffice; to raise total output the capacity to produce must be increased. In terms of this model K must grow. From equation (3) if we assume that s and v are constants, it is clear that I must also grow. Rearranging equation (3) we know that

$$I = \frac{s}{v}(K) \qquad (3a)$$

i.e. that this is the level of investment that fully uses the savings willingly made out of full capacity income. Dividing both sides of the equation by K,

$$\frac{I}{K} = \frac{s}{v} \qquad (3b)$$

This is even more illuminating. Since I represents the *net addition* to the stock of capital in any given period and K represents the stock of capital in that same period, $\frac{I}{K}$ is the rate of growth of the stock of capital. This rate should be $\frac{s}{v}$ per annum.

We have now established the conditions for an equilibrium rate of growth in the economy. If the capital stock grows at a rate of $\frac{s}{v}$ then the extra output generated will just be absorbed by the increase in total expenditure. It can in fact also be shown that not only the capital stock, but investment and income too must grow at a rate of $\frac{s}{v}$ per annum for equilibrium to be preserved.

3. Effect of a Fall in Primary Product Exports on Producing Countries' Incomes

A decline in prices and so the value of primary products in world markets reduces the incomes of the producers of these commodities. A reduction in incomes in "poor" countries leads to a reduction in expenditure, both on domestically-produced goods and on imports. The reduction in import spending by "poor" countries reduces the export proceeds, hence incomes in the rest of the world (the "rich" countries), and in consequence *their* spending on their own domestic goods and on imports from the "poor" countries declines, which in turn has further repercussions. There is a continuing chain of reactions, as exports and incomes fall first in one sector of the world, then another.

We wish to ascertain what the full repercussions of any initial (exogenous) disturbance are likely to be on levels of income in "rich" and "poor" countries. Given the complicated nature of the problem—the inter-acting secondary and tertiary repercussions, and so forth—it is difficult to reason the problem through verbally, and so we must resort to elementary mathematics. The most difficult problem in using elementary mathematics to solve economic problems usually lies in seeing how to *set the problem up*—the mathematics may, as in this case, mean simply solution of two linear simultaneous equations.

Let us denote a change in the income of the "poor" countries by $\triangle Y_1$, and a change in the income of the "rich" countries by $\triangle Y_2$. We assume that income in both "poor" and "rich" countries is disposed of in three ways: it is spent on domestically produced goods, it is spent on imported goods, or it is saved. To know the full consequences of any change in income we need to know the marginal propensities to buy domestically produced goods, to import and to save. Suppose past statistical evidence suggested the following values:

For "poor" countries	*For "rich" countries*
$d_1 = \cdot 7$	$d_2 = \cdot 6$
$m_1 = \cdot 2$	$m_2 = \cdot 3$
$s_1 = \cdot 1$	$s_2 = \cdot 1$

Hence changes in income in the two areas will lead to the following changes:

	$\triangle Y_1$	$\triangle Y_2$
Changes in domestic spending	$d_1\triangle Y_1 = \cdot 7\triangle Y_1$	$d_2\triangle Y_2 = \cdot 6\triangle Y_2$
Changes in import spending	$m_1\triangle Y_1 = \cdot 2\triangle Y_1$	$m_2\triangle Y_2 = \cdot 3\triangle Y_2$
Changes in saving	$s_1\triangle Y_1 = \cdot 1\triangle Y_1$	$s_2\triangle Y_2 = \cdot 1\triangle Y_2$

The income of the two areas, however, is determined by the value of their total production. This will include goods produced for domestic use, and goods produced for export. We have assumed that the exports of the "poor" countries' are the imports of the "rich" countries, and vice versa. Changes in the total product (which is equivalent to the total income) may therefore be expressed as follows:

$$\triangle Y_1 = d_1 \triangle Y_1 + m_2 \triangle Y_2$$
$$\triangle Y_2 = d_2 \triangle Y_2 + m_1 \triangle Y_1$$

or using the values given

$$\triangle Y_1 = \cdot 7 \triangle Y_1 + \cdot 3 \triangle Y_2$$
$$\triangle Y_2 = \cdot 6 \triangle Y_2 + \cdot 2 \triangle Y_1$$

The problem before us is to discover the full repercussions in terms of falling incomes ($\triangle Y_1$ and $\triangle Y_2$) of an autonomous drop in the value of the "poor" countries' exports. Let the drop in value be $-a$; in the initial period only the incomes of the "poor" countries are affected—"rich" countries are so far unaffected.

$$\triangle Y_1 = \cdot 7 \triangle Y_1 + \cdot 3 \triangle Y_2 - a \quad (1)$$
$$\triangle Y_2 = \cdot 6 \triangle Y_2 + \cdot 2 \triangle Y_1 \quad (2)$$

The difficult part of the task is now completed. Having set the problem up, we can proceed to solve for $(\triangle Y_1)$ and $(\triangle Y_2)$ in terms of the initial disturbance $(- a)$. Rearranging the equations we have:

$$\triangle Y_1 - \cdot 7(\triangle Y_1) - \cdot 3(\triangle Y_2) = - a, \text{ or}$$
$$\cdot 3(\triangle Y_1) - \cdot 3(\triangle Y_2) = - a, \quad (1)$$

and

$$\triangle Y_2 - \cdot 6(\triangle Y_2) - \cdot 2(\triangle Y_1) = 0, \text{ or}$$
$$\cdot 4(\triangle Y_2) - \cdot 2(\triangle Y_1) = 0. \ (2)$$

Thus we have

$$\cdot 3(\triangle Y_1) - \cdot 3(\triangle Y_2) = - a \quad (1)$$
$$- \cdot 2(\triangle Y_1) + \cdot 4(\triangle Y_2) = 0 \quad (2)$$

Multiplying the first equation by 4 and the second by 3 and adding (to eliminate $\triangle Y_2$):

$$1\cdot 2(\triangle Y_1) - 1\cdot 2(\triangle Y_2) = -4a \quad (1) \times 4$$
$$- \cdot 6(\triangle Y_1) + 1\cdot 2(\triangle Y_2) = 0 \quad (2) \times 3$$

$$\cdot 6(\triangle Y_1) = -4a$$
$$\triangle Y_1 = -6\cdot 67a$$

And substituting this value of $\triangle Y_1$ in the second equation

$$\triangle Y_2 = \cdot 6\triangle Y_2 + \cdot 2\triangle Y_1 \ (2)$$
$$\cdot 4\triangle Y_2 = \cdot 2\,(-6\cdot 67a)$$
$$\triangle Y_2 = -3\cdot 33a.$$

Every *initial* £1 fall in export proceeds of "poor" countries will produce a *total* decline of £6·67 in incomes in "poor" countries and a *total* decline of £3·33 in incomes in "rich" countries. If the initial fall amounted to £10 million, incomes in "poor" countries would decline by £67 million, incomes in "rich" countries would decline by £33 million.

EXERCISE 5.

SECTION A. PRICE THEORY

1. Would the following equations be more suitable for representing supply or demand equations:

$$Q = 155 - 25P$$
$$Q = 50P$$
$$P = \cdot 10Q^2$$
$$PQ = 20$$
$$5Q - 50 - 200P = 0$$
$$Q = 1200 - P^2$$

What assumptions have you made about the characteristic shape of supply and demand functions?

2. Complete the models and find the equilibrium price and quantity traded for, given the following equations:

$$\text{(a)}\quad Q^d = 100 - 20P$$
$$Q^s = -5 + 15P$$
$$\text{(b)}\quad Q^d = 1000 - P^2$$
$$Q^s = 30P$$

Graph the two market situations.

3. Suppose that a demand equation is given by

$$Q^d = 50 - 3P + Y$$

where Y represents consumer's income, and a supply equation by

$$Q^s = \frac{P^2}{2}$$

Find the equilibrium price and quantity
(a) when $Y = 12$
(b) when $Y = 60$

4. It is known in the University Bookshop that when the price of a certain paperback book is ten shillings there will be no purchases, but it is thought that for every shilling drop in price, twelve new purchasers will appear. Furthermore, it is known that the publishers refuse to offer any copies for sale at a price of three shillings, or less; but they are prepared to offer nine copies for four shillings and an additional nine copies for each one shilling rise in price thereafter.
(a) Find two equations showing
    (i) the quantity demanded (Q^d) in terms of price (P):
    (ii) the quantity supplied (Q^s) in terms of price (P).
(b) Find the "equilibrium" price at which the quantity demanded will equal the quantity supplied, and the number of books the publisher must supply.

5. (This question is about supply and demand, but it also requires knowledge about the behaviour of a firm in circumstances of perfect competition.)

Bread is being produced in two bakeries in Kampala. They compete keenly and there is no collusion between them—explicit or implicit. The bread they produce is identical.
x_1 represents the output of the first bakery (in dozens of loaves)
x_2 represents the output of the second bakery (in dozens of loaves).
The marginal cost in cents of the firms is described by

$$MC_1 = 400 + \tfrac{1}{2}x_1$$
$$MC_2 = 2x_2 + 100.$$

The demand for bread in Kampala is 4,500 dozen loaves a day (at any price).

(a) What equation is needed to complete the model?

(b) How many loaves of bread will be produced by
   (i) Bakery No. 1
   (ii) Bakery No. 2?

(c) At what price(s) will bread be sold in the two bakeries?

6. A consumer has an income of Shs. 2000, which he can spend either on good X or on good Y. The price of X is Shs. 50 per unit, the price of Y is Shs. 25 per unit.

(a) Express, in the form of an equation, the constraint on this consumer's expenditure. The general form of the equation will be $E = f(X,Y)$.

(b) Draw the graph of this constraint line.

(c) Explain the economic meaning of the *slope* of this function.

(d) Graph two further budget constraints, showing the effect of
   (i) an increase in the price of Y to Shs. 40 per unit;
   (ii) an increase in the consumer's income to Shs. 2500. (Assume prices remain Shs. 50 for X and Shs. 25 for Y).

SECTION B. INCOME THEORY

1. The economy of a country may be described by the following equations:

$$C = 15 + \cdot 9Y$$
$$I = 20 + \cdot 05Y$$
$$G = 25$$

(a) Complete the model and solve for the equilibrium values of Y, C, I and G.

(b) Draw an accurate graph of the three functions C, C + 1, and C + 1 + G and give the equilibrium levels of income with and without Government expenditure.

2. The national income of Arcadia may be described by the following equations:

$$S = \cdot 25Y - 100$$
$$I = 250$$

(a) Complete this model and find the equilibrium level of income and savings.

(b) Graph the above information on a diagram.

(c) Show by algebra and by graph what happens to the savings function and the equilibrium level of income when
   (i) People decide to save an additional 75 at all levels of income.
   (ii) People decide to save 50 less at all levels of income.

3. An urban worker in Kampala is obliged to spend Shs. 110 each month on the necessities of life. Surveys suggest that each extra ten shillings earned make it possible for him to save Shs. 1·50 a month. What is the equation of the worker's consumption function?

4. (a) Explain clearly the relation between the consumption function and the savings function for an economy. If the consumption function is

$$C = 30 + \cdot 8Y$$

What is the equation for savings? Plot these two functions on the same diagram.

(b) Now assume that investment is equal to 20, and show graphically that the equilibrium level of C + I is at the point where savings and investment are equal.

5. (a) Will the equilibrium level of income in an economy necessarily be one of full employment? Why not?

(b) If a country has a potential "full-employment" income of 1000 and the following consumption and investment functions

$$C = \cdot 75Y + 50$$
$$I = 150,$$

find the level of Government expenditure necessary to maintain full employment.

(c) What is the value of the "multiplier" in this economy?

6. (a) Let I_t, K_t and Y_t represent investment, capital stock and income at time t. I_{t+1} K_{t+1} and Y_{t+1} represent the same variables at time t+1. s represents the marginal and average propensity to save. v represents the marginal and average capital/output ratio. Given the equilibrium conditions

$$I_t = \frac{s}{v} K_t$$

$$I_{t+1} = \frac{s}{v} K_{t+1}$$

and the fact that

$$K_{t+1} = K_t + I_t,$$

prove that investment must grow at a rate of $\frac{s}{v}$

(Suggestion: rate of growth of investment $= \frac{I_{t+1} - I_t}{I_t}$

(b) Given the equilibrium rate of growth of investment

$$\frac{I_{t+1} - I_t}{I_t} = \frac{s}{v}$$

and the equilibrium conditions

$$I_t = sY_t$$
$$I_{t+1} = sY_{t+1}$$

show that national income (Y) will also grow at a rate of $\frac{s}{v}$

(c) If $I_t = 300$, $s = 0{\cdot}15$, $v = 3$:
(i) What should be I_{t+1}?
(ii) What is the rate of growth of investment?
(iii) What is the equilibrium level of income at time (t) and time (t+1)?
(iv) What is the rate of growth of income?

7. Obtain a general (literal) solution for the change in incomes in "rich" and "poor" countries when there is an autonomous fall in the value of the former's primary product purchases from the latter.

8. How would you set up the two-country income model if the initial disturbance "a" was a *switch* in spending by the "poor" countries from imports to home goods?

PART TWO
INTRODUCTION TO CALCULUS

VI. DIFFERENTIATION

A. SLOPES AND LIMITS

1. Rates of Change and Gradients

(i) *Linear functions*

We may introduce the concept of a slope or gradient by looking at the graphs of two linear functions:

$y = 3x + 4$

x	–1	0	1	2	3
y	1	4	7	10	13

$y = 3x - 1$

x	–1	0	1	2	3
y	–4	–1	2	5	8

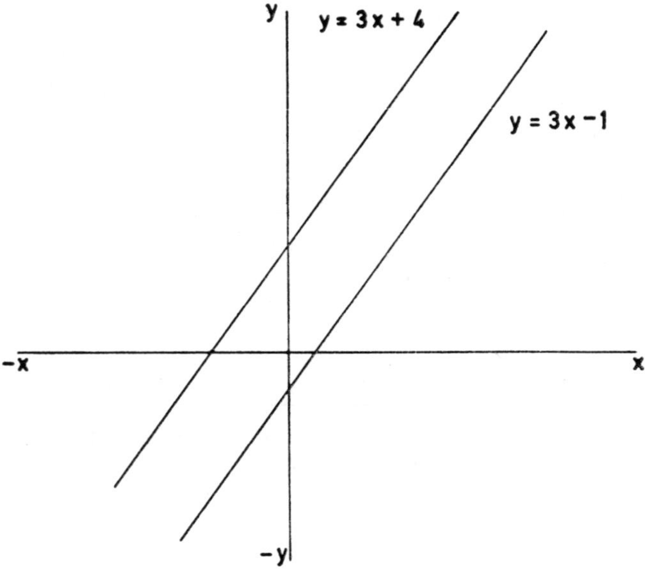

These two graphs are parallel: they have the same slope. The reason for the identical slope of the two graphs is that the rate of increase of y is related to the rate of increase of x, in exactly the same way in both cases. This is evident from the tables: every increase of 1 in x is accompanied by an increase of 3 in y.

The ratio of the increase in y to the increase in x measures the *rate of change* of the function $y = f(x)$. The mathematical symbol for "an increase or decrease in" a variable is $\triangle$. Thus the rate of change of the above linear functions is measured by $\frac{\triangle y}{\triangle x}$

In graphical terms $\frac{\Delta y}{\Delta x}$ measures the *gradient* of the line between two points.

For both the above functions, $y = 3x + 4$ and $y = 3x - 1$,

$$\frac{\Delta y}{\Delta x} = 3,$$

and for any linear function of the form $y = ax + b$,

$$\frac{\Delta y}{\Delta x} = a.$$

The slope of a straight line, as measured by the ratio $\frac{\Delta y}{\Delta x}$ may be positive or negative. If $\frac{\Delta y}{\Delta x}$ is positive then y increases as x increases and the graph of the function slopes upwards to the right. If $\frac{\Delta y}{\Delta x}$ is negative, then y decreases as x increases and the graph of the function slopes downwards to the right.

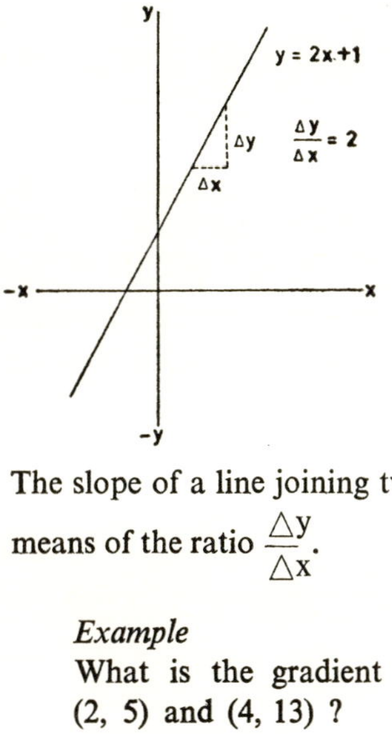

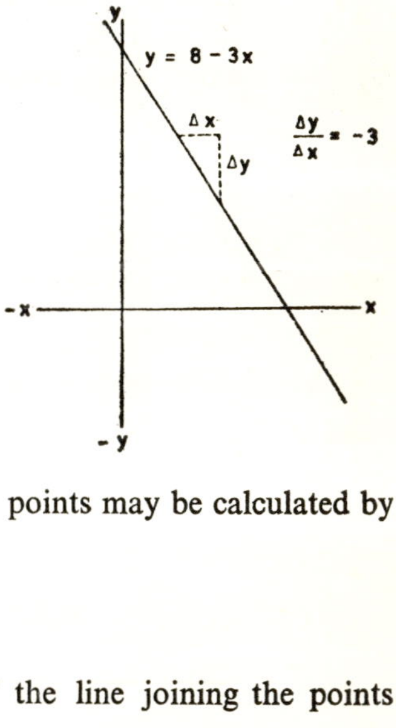

The slope of a line joining two points may be calculated by means of the ratio $\frac{\Delta y}{\Delta x}$.

Example

What is the gradient of the line joining the points (2, 5) and (4, 13) ?

Solution

$$\text{Gradient} = \frac{\Delta y}{\Delta x} = \frac{13 - 5}{4 - 2} = 4$$

(ii) *A note on Tangents*

The slope of a straight line is of course closely related to the tangent of an angle. In the triangle ABC with a 90° angle at **C** the tangent of the angle at **B** (angle θ) is

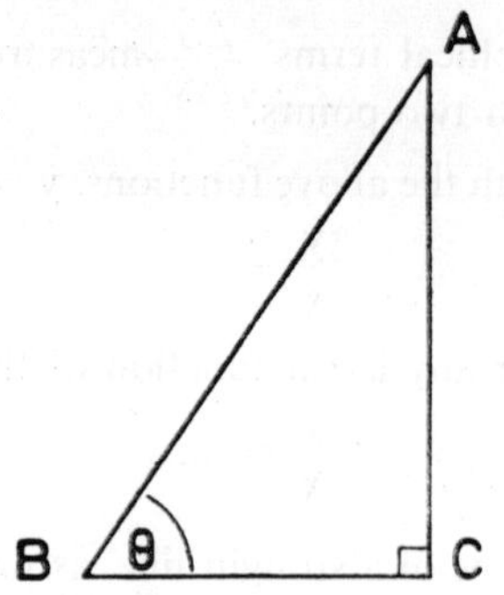

measured by the ratio of the opposite side divided by the adjacent side

$$\tan \theta = \frac{AC}{BC}$$

This ratio which measures the tangent of angle θ also measures the slope of AB. (Inspection shows that the size of angle θ determines the slope of AB).
If this triangle were drawn on a graph, with the points A, B and C given by the coordinates (5, 7), (2,1) and (5, 1)

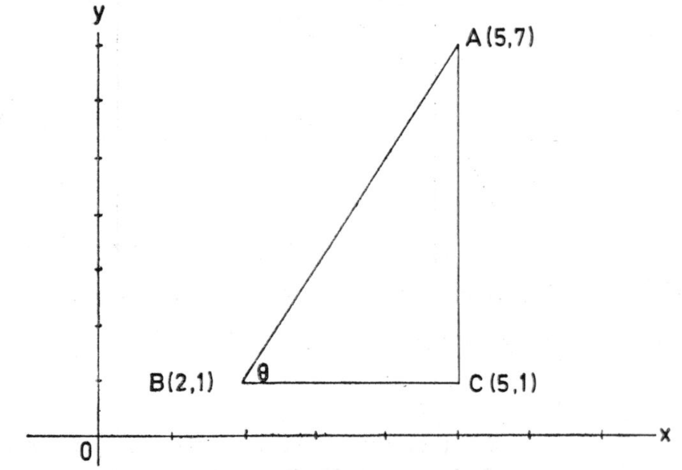

$$\text{the tangent of angle } \theta = \frac{AC}{AB} = \frac{7 - 1}{5 - 2} = \frac{3}{1}$$

As A and C have the same x coordinates and B and C have the same y coordinates, this ratio could be expressed as the difference of the y coordinates between A and B divided by the difference of the x coordinates between A and B. This measurement of the tangent of angle θ coincides exactly with our definition of a *slope*.

(iii) *Non-linear Functions*

A straight line has the same slope throughout; the graphs of all other functions have different slopes at different sections of the curve.

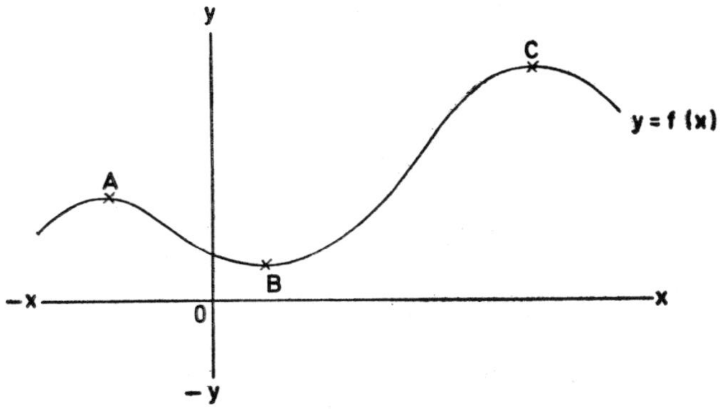

The above curve has a different slope at every point. The gradient of the curve as measured by

$$\frac{\text{rate of increase of y}}{\text{rate of increase of x}} = \frac{\Delta y}{\Delta x}$$

is positive to the left of A, negative from A to B, and positive again from B to C.
The problem now is to calculate the slope of the curve at any particular point. This could be done if the equation of the tangent to the curve at that point were known.

Example

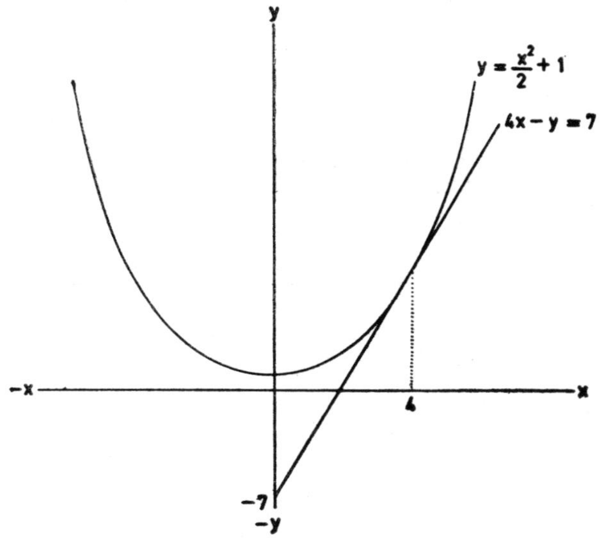

The straight line $4x - y = 7$ is tangential to the curve $y = \frac{x^2}{2} + 1$, at the point (4, 9). What is the slope of the curve at this point?

Solution

The tangent to the curve when written in the form $y = ax + b$ is $y = 4x + 7$. The gradient of this straight line is

$$\frac{\Delta y}{\Delta x} = 4$$

At the point (4, 9) the curve and the tangent have the same slope, therefore at (4, 9) the slope of $y = \frac{x^2}{2} + 1$ is also equal to 4.

If, however, the tangent to the curve at the point is not known, then we have no method of measuring the gradient of a non-linear curve. Calculus provides the necessary techniques.

2. Limits

(i) The concept of a limit may be introduced by means of a problem.

The speed of a train can be calculated if two other variables are known—distance and time.

$$S = \frac{D}{T}$$

How could a man standing at point A on a railway line calculate the speed of a passing train *at that point*?

If he knew that the train reached B, 500 feet away in 10 seconds, then he could calculate the speed of the train from A to B.

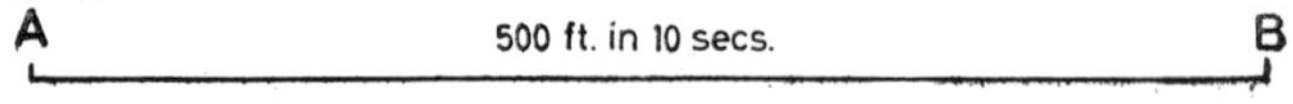

$$\text{Speed from A to B} = \frac{500}{10} = 50 \text{ ft./sec.}$$

If the train reaches C, 288 feet from A in 6 seconds,

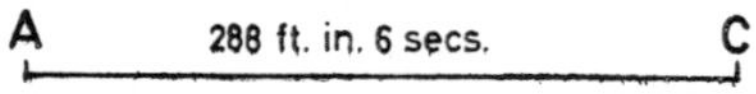

Speed from A to C $= \frac{288}{6} = 48$ ft./sec.

If the train reaches D, 143 feet away from A, in 3 seconds, then

A 143 ft. in 3 secs. D

Speed from A to D $= \frac{143}{3} = 47 \cdot 6$ ft./sec.

If the train reaches E, 95 feet from A, in 2 seconds, then

A 95 ft. in 2 secs. E

Speed from A to E $= \frac{95}{2} = 47 \cdot 5$ ft./sec.

In this example, we have been taking points nearer and nearer A along the railway line, and calculating the speed from A to the different points. The results are:

A to B	50 ft./sec.
A to C	48 ft./sec.
A to D	47·6 ft./sec.
A to E	47·5 ft./sec.

It is still impossible to calculate the exact speed at A where both distance and time are equal to 0, but it would be possible to take further points nearer and nearer to A, and make further calculations of speed, and it seems likely that the speed at A is around 47·5 ft./sec.

This method of calculating the speed at A by calculating $\frac{D}{T}$ for various distances and times, each a little smaller than the last, suggests that the speed at A may be thought of as the *limit* of D/T as D and T approach 0.

(ii) In algebra it is often useful to calculate the limit of f(x) as x becomes smaller and smaller and approaches 0, or in mathematical terms.

$$x \longrightarrow 0$$

Examples

(a) $f(x) = 3x^2 + x + 3$

As $x \longrightarrow 0$,

$$f(x) \longrightarrow 3(0)^2 + (0) + 3$$
$$= 3$$

The limit of the function as $x \rightarrow 0$ is 3.

(b) $f(x) = \dfrac{3x^2 - 4}{2x + 1}$

As $x \longrightarrow 0$

$f(x) \longrightarrow \dfrac{3(0)^2 - 4}{2(0) + 1}$

$= -4$

The limit of the function as $x \to 0$ is -4.

B. DIFFERENTIATION FROM FIRST PRINCIPLES

1. A Numerical Example

$y = x^2$ is a quadratic function of x. At any point lying on the graph of the function, the y-coordinate will be equal to the square of the x-coordinate.

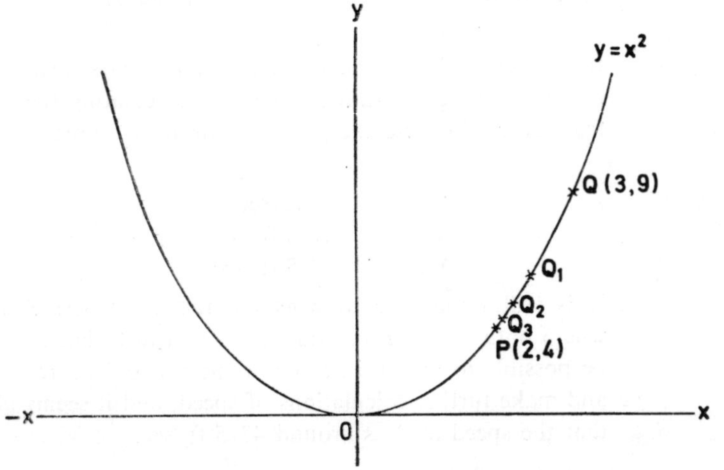

Let P(2,4) and Q(3,9) be points on the curve.
The problem is to calculate the *slope at P*, and the method is first to calculate the slope of PQ.

$$\text{Slope of PQ} = \frac{\Delta y}{\Delta x} = \frac{9 - 4}{3 - 2} = \frac{5}{1} = 5$$

Now, as in the train example, choose points Q_1, Q_2, Q_3, ... nearer and nearer to P.

Let Q_1 have the coordinates (2·5,6·25)

$$\text{Slope of } PQ_1 = \frac{6{\cdot}25 - 4}{2{\cdot}5 - 2} = 4{\cdot}5$$

Let Q_2 have coordinates (2·3,5·29)

$$\text{Slope of } PQ_2 = \frac{5{\cdot}29 - 4}{2{\cdot}3 - 2} = 4{\cdot}3$$

Let Q_3 have coordinates (2·1,4·41)

$$\text{Slope of } PQ_3 = \frac{4\cdot41 - 4}{2\cdot1 - 2} = 4\cdot1$$

Q_3 is very near to P and it seems reasonable to assume that as Q approaches P, the slope tends to the limit of 4. This conclusion, however, is only an intelligent guess, and to reach a more exact solution of the problem it is necessary to turn to algebra.

2. An Algebraic Example

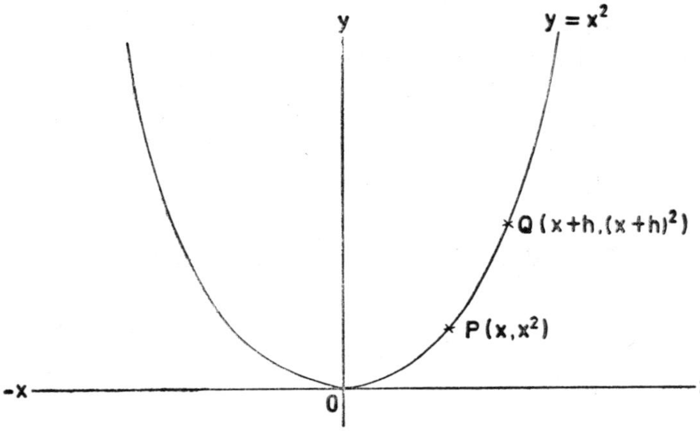

This time, let P have the coordinates (x,x^2) let Q have coordinates $(x + h, (x + h)^2)$ (h represents a small, unspecified, increment in x. The y-coordinate is, of course, always equal to the square of the x-coordinate on $y = x^2$).

Again, to calculate the slope *at* P, consider first the slope of PQ:

$$\text{Slope PQ} = \frac{\Delta y}{\Delta x} = \frac{(x + h)^2 - x^2}{(x + h) - x}$$

$$= \frac{x^2 + 2xh + h^2 - x^2}{x + h - x}$$

$$= \frac{2xh + h^2}{h}$$

$$= 2x + h.$$

As Q approaches P, $h \longrightarrow 0$, and the slope $\dfrac{\Delta y}{\Delta x}$ tends to the limit of 2x.

> (Compare this result with that of the numerical example, There it seemed that the limit of the slope at P(2,4) was 4. or twice the x-coordinate 2.)

This result, that the slope $\dfrac{\Delta y}{\Delta x}$ tends to a limit of 2x is true for any point P (x,x^2) on the curve $y = x^2$.

Example

At (3,9) $\frac{\Delta y}{\Delta x} \rightarrow 2x = 2(3) = 6$

At (– 4,16) $\frac{\Delta y}{\Delta x} \rightarrow 2x = 2(-4) = -8$

(i.e. the curve is dropping sharply at (– 4,16).

3. Standard Presentation

(i) The calculation of the rate of change of a function is known as the process of *differentiation.*

The method of differentiation that follows uses the same ideas as the preceding algebraic example, but presents them in a different form. The student who can think his way through both methods should by the end have a clear understanding of what is meant by "differentiation".

(ii) *An Illustration*

To find the gradient at any point of the curve $y = x^2$

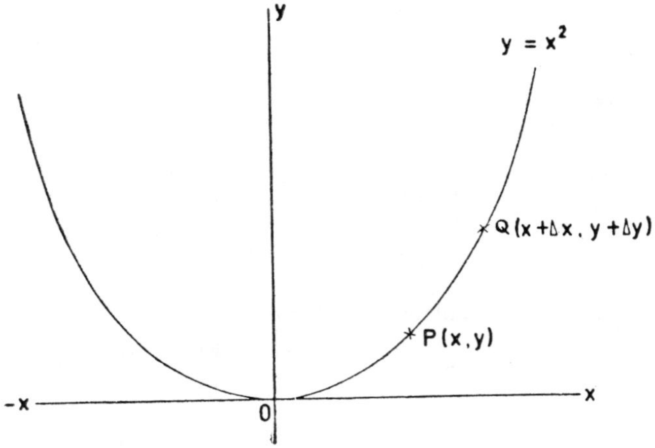

Let P(x,y) and Q(x + Δx, y + Δy) be two points on the curve (Δx, Δy, represent small increases of x and y, and as the gradient of the curve is equal to the change in y divided by the change in x, we are trying to calculate $\frac{\Delta x}{\Delta y}$.)

At Q $\quad y + \Delta y = (x + \Delta x)^2$
$\quad = x^2 + 2x\Delta x + (\Delta x)^2$

At P $\quad y = x^2$

$\Delta y = 2x\Delta x + (\Delta x)^2$ (by subtraction)

$\frac{\Delta y}{\Delta x} = \frac{2x\Delta x}{\Delta x} + \frac{(\Delta x)^2}{\Delta x}$ (divide by Δx)

$= 2x + \Delta x$

As Q → P

$$\Delta x \to 0$$
$$\frac{\Delta y}{\Delta x} \to 2x.$$

The name of the limit of $\frac{\Delta y}{\Delta x}$ as $\Delta x \to 0$ is $\frac{dy}{dx}$, or the *derivative* of the function; therefore the final solution may be written

$$\frac{dy}{dx} = 2x$$

(iii) *Another Illustration*

To find the gradient at any point of the curve $y = 2x^2 + 3$

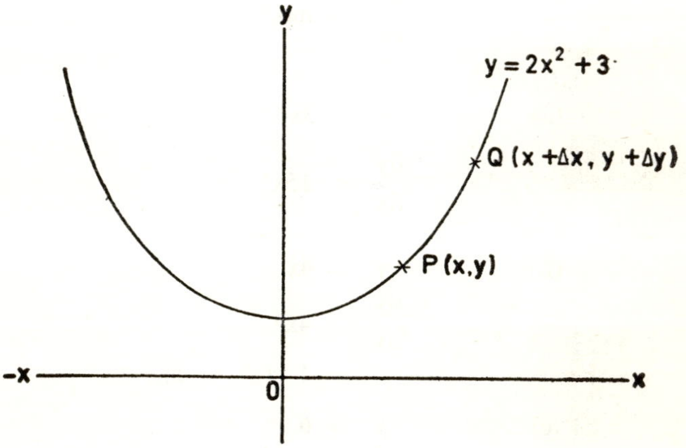

Let P(x,y) and Q(x + Δx, y + Δy) be two points on the curve.

At Q $\quad y + \Delta y = 2(x + \Delta x)^2 + 3$
$$= 2(x^2 + 2x\Delta x + (\Delta x)^2) + 3$$
$$= 2x^2 + 4x\Delta x + 2(\Delta x)^2 + 3$$

At P $\quad y = 2x^2 + 3$

$$\Delta y = 4x\Delta x + 2(\Delta x)^2$$

$$\frac{\Delta y}{\Delta x} = \frac{4x\Delta x}{\Delta x} + \frac{2(\Delta x)^2}{\Delta x}$$
$$= 4x + 2\Delta x$$

As Q → P

$$\Delta x \to 0$$

$$\frac{\Delta y}{\Delta x} \to \frac{dy}{dx} = 4x$$

C. TECHNIQUES OF DIFFERENTIATION

The derivative of any function $y = f(x)$ measures the rate of change of the function at any point. The derivative is defined as

$$\frac{dy}{dx} = \text{the limit of } \frac{\triangle y}{\triangle x} \text{ as } \triangle x, \triangle y, \text{ tend to } 0.$$

The method of differentiation from first principles should be understood, but fortunately there are a number of short-cut methods which can normally be used. A list of these rules for differentiation follows.

1. Standard Form

$$y = ax^n$$

$$\frac{dy}{dx} = nax^{n-1}.$$

Examples

(a)

$$y = 3x^5$$

$$\frac{dy}{dx} = 15x^4$$

(b)

$$y = 4x$$

$$\frac{dy}{dx} = 4x^0$$

$$= 4 \quad (x^0 = 1)$$

(c)

$$y = 6$$

This could be written

$$y = 6x^0 \quad (\text{since } x^0 = 1)$$

$$\frac{dy}{dx} = 0 \cdot 6x^{-1}$$

$$= 0$$

2. Two Special Cases of the Standard Form

(i) *When n is negative*

By definition (see Section 1.1)

$$\frac{1}{x^n} = x^{-n}$$

Therefore to differentiate

$$y = \frac{1}{x^2} = x^{-2}$$

$$\frac{dy}{dx} = -\ 2x^{(-2-1)} = -2x^{-3}$$

$$= -\frac{2}{x^3}$$ (It is customary not to leave negative indices in answers)

(ii) *When n is fractional*

By definition (see Section 1.1)

$$\sqrt[n]{x} = x^{\frac{1}{n}}$$

$$\sqrt[n]{x^p} = x^{\frac{p}{n}}.$$

Therefore to differentiate

$$y = \sqrt{x} = x^{\frac{1}{2}}$$

$$\frac{dy}{dx} = \frac{1}{2}x^{(\frac{1}{2}-1)}$$

$$= \frac{1}{2}x^{-\frac{1}{2}}$$

$$= \frac{1}{2\sqrt{x}}$$

Further Examples

(a) $$y = \frac{1}{x^3} = x^{-3}$$

$$\frac{dy}{dx} = -\ 3x^{-4}$$

$$= \frac{-3}{x^4}$$

(b) $$y = \frac{3}{x^5} = 3x^{-5}$$

$$\frac{dy}{dx} = -\ 15x^{-6}$$

$$= \frac{-15}{x^6}$$

(c) $$y = \sqrt[3]{x} = x^{\frac{1}{3}}$$

$$\frac{dy}{dx} = \frac{1}{3}x^{-\frac{2}{3}}$$

$$= \frac{1}{3\sqrt[3]{x^2}}$$

3. The Derivative of a Sum or Difference

If $y = u + v$ (where u and v are functions of x)

$$\frac{dy}{dx} = \frac{du}{dx} + \frac{dv}{dx}$$

Example

$$y = 3x^4 - 4x^2$$

$$\frac{dy}{dx} = 12x^3 - 8x$$

4. The Derivative of a Product

If $y = uv$

$$\frac{dy}{dx} = v\frac{du}{dx} + u\frac{dv}{dx}$$

Example

$$y = (2x + 1)(3x^2) \qquad \left[u = 2x + 1, \frac{du}{dx} = 2 \quad v = 3x^2, \frac{dv}{dx} = 6x\right]$$

$$\frac{dy}{dx} = 2(3x^2) + 6x)(2x + 1)$$

$$= 6x^2 + 12x^2 + 6x$$

$$= 18x^2 + 6x.$$

5. The Derivative of a Quotient

$$y = \frac{u}{v}$$

$$\frac{dv}{dx} = \frac{v\frac{du}{dx} - u\frac{dv}{dx}}{v^2}.$$

Example

$$y = \frac{7x - 2}{5x + 3} \qquad \left[u = 7x - 2, \frac{du}{dx} = 7 \quad v = 5x + 3, \frac{dv}{dx} = 5\right]$$

$$\frac{dy}{dx} = \frac{7(5x + 3) - 5(7x - 2)}{(5x + 3)^2}$$

$$= \frac{35x + 21 - 35x + 10}{(5x + 3)^2}$$

$$= \frac{31}{(5x + 3)^2}.$$

6. The Derivative of a Function of a Function

If $y = f(u)$ where $u = f(x)$

$$\frac{dy}{dx} = \frac{dy}{du} \cdot \frac{du}{dx}$$

Example

$$y = (3x^2 + 2)^5 \qquad \left[y = u^5, \ \frac{dy}{du} = 5u^4 \right.$$

$$\left. u = 3x^2 + 2, \ \frac{du}{dx} = 6x \right]$$

$$\frac{dy}{dx} = 5u^4 . \ 6x$$
$$= 30x \ . \ u^4$$
$$= 30x \ (3x^2 + 2)^4$$

D. USE OF THE DERIVATIVE IN A NUMERICAL CALCULATION

The notation $\frac{dy}{dx}$ is the derivative of the function $y = f(x)$ and is the general expression for calculating the rate of change ot the function. $\frac{dy}{dx}$ was defined as the limit of $\frac{\triangle y}{\triangle x}$ as $\triangle y$ and $\triangle x$ approached 0. For small values of x and y it is therefore *approximately* true that

$$\frac{\triangle y}{\triangle x} \approx \frac{dy}{dx}$$

$$\text{or} \quad \triangle y \approx \frac{dy}{dx} (\triangle x)$$

This approximation can be used in problems involving the rates of change of two variables.

Example
The point (4,8) lies on the curve $y = x^2 - 5x + 12$. What is the increase in y as x increases from 4 to 4·1 ?

$$y = x^2 - 5x - 12$$

$$\frac{dy}{dx} = 2x - 5$$

$$\triangle y \approx \frac{dy}{dx} (\triangle x)$$

$$\approx (2x - 5)(\cdot 1)$$

$$\approx (2(4) - 5)(\cdot 1)$$

$$\approx (3)(\cdot 1)$$

$$\approx \cdot 3$$

Answer

As x increases from 4 to 4.1, y increases by approximately ·3, from 8 to 8.3.

Note

This solution may be checked by substitution of 4.1 for x in the original function:

$$\begin{aligned} y &= (4\cdot 1)^2 - 5(4\cdot 1) - 12 \\ &= 16\cdot 81 - 20\cdot 5 - 12 \\ &= 8\cdot 31. \end{aligned}$$

EXERCISE 6.

A. Differentiate from first principles; and make a rough sketch of each curve:

1. $y = 3x - 5$
2. $y = 2x^2 + 1$
3. $y = x^3$
4. $y = x^2 - 2x$

B. Differentiate by rule:

1. $y = x^7$
2. $y = 3x^2 + 2x$
3. $y = 5x^2 - 3x + 4$
4. $y = 6x^3 + \frac{x^2}{2} - x + 1$
5. $y = \frac{1}{x^4}$
6. $y = \frac{3}{x^4}$
7. $y = \sqrt{x}$

8. $y = \dfrac{1}{\sqrt{x}}$

9. $y = 6x^{\frac{1}{3}}$

10 $y = (3x + 2)(x^2 + 1)$

11. $y = (1 - 3x)(3 + x^2)$

12. $y = (x^3 + 5)(2x^2 - 4)$

13. $y = \dfrac{x^2 + 2}{4x - 1}$

14. $y = \dfrac{3x - 4}{x^2 + 3}$

15. $y = \dfrac{6 - x}{2 - 5x^2}$

16. $y = (2x - 7)^3$

17 $y = (5 - x^2)^7$

18. $y = (x^2 - 3x)^5$

19. $y = ax + b$

20. $y = ax^2 + bx + c$

21. $y = ax^3 + bx^3 + cx + d$

C. 1. Calculate the approximate change in y on the function $y = x^2 + x - 2$ as x increases from 2 to 2·1.

2. Calculate the approximate change in y on the function $y = 12 + 8x - x^2$ as x increases from 3 to 3.1.

VII. MAXIMA, MINIMA AND POINTS OF INFLEXION

1. Maximum and Minimum Turning Points

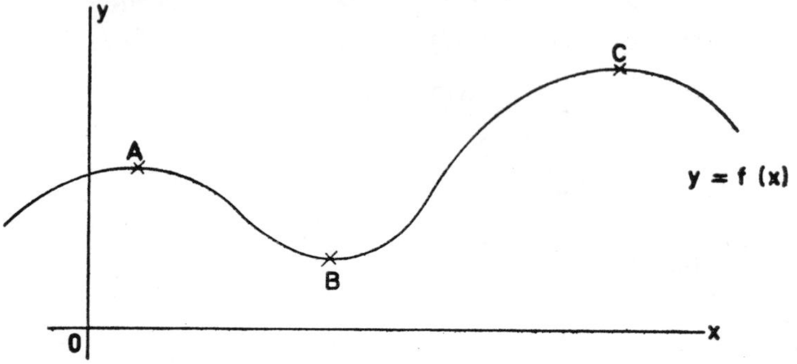

In the above diagram A, B, and C represent turning points on the curve y = f(x).

A and C are maximum turning points
B is a minimum turning point.

The gradient of the curve, measured by $\frac{dy}{dx}$ is equal to 0 at all turning points.

At a *maximum turning point* the sign of $\frac{dy}{dx}$ changes from positive through 0 to negative as x increases.

At a *minimum turning point* the sign of $\frac{dy}{dx}$ changes from negative through 0 to positive as x increases.

Example

$$y = x^2 - 6x - 3$$

Find the turning point of the above curve and show whether it is a maximum or a minimum.

$$y = x^2 - 6x - 3$$

$$\frac{dy}{dx} = 2x - 6$$

At a turning point $\frac{dy}{dx}$ is equal to 0

Let $\frac{dy}{dx} = 0$

$$2x - 6 = 0$$
$$x = 3$$

When $x = 3$, $y = 3^2 - 6(3) - 3$
$= 9 - 18 - 3$
$= -12$

There is a turning point at $(3, -12)$

What is happening to the sign of $\frac{dy}{dx}$ just before and just after the point where $x = 3$?

Let $x = 2{\cdot}9$

$$\frac{dy}{dx} = 2(2{\cdot}9) - 6$$
$$= 5{\cdot}8 - 6$$
$$= -{\cdot}2 \text{ (i.e. } \frac{dy}{dx} \text{ is negative)}$$

Let $x = 3.1$

$$\frac{dy}{dx} = 2(3{\cdot}1) - 6$$
$$= 6{\cdot}2 - 6$$
$$= {\cdot}2 \quad \text{(i.e. } \frac{dy}{dx} \text{ is positive)}$$

Therefore at the point $(3, -12)$ $\frac{dy}{dx}$ is changing from negative through 0 to positive and there is a *minimum* turning point.

2. Points of Inflexion

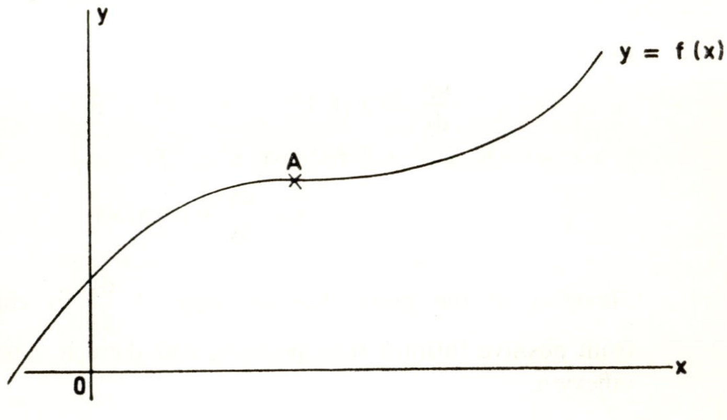

The above curve $y = f(x)$ has at A a value of x for which $\frac{dy}{dx} = 0$.

The sign of $\frac{dy}{dx}$ at A is changing from positive through 0 to positive again, so this is neither a maximum nor a minimum turning point, but a *point of inflexion.*

Example

Find the turning points and/or points of inflexion for the curve

$$y = x^3 - 3x^2 + 3x + 2$$

We have

$$y = x^3 - 3x^2 + 3x + 2$$

$$\frac{dy}{dx} = 3x^2 - 6x + 3$$

Let $\frac{dy}{dx} = 0$

$$3x^2 - 6x + 3 = 0$$

$$x^2 - 2x + 1 = 0$$

$$(x - 1)(x - 1) = 0$$

$$x = 1$$

When $x = 1$, $y = (1)^3 - 3(1)^2 + 3(1) + 2$

$$= 1 - 3 + 3 + 2$$

$$= 3$$

There is a *stationary* value at (1,3).

Let $x = \cdot 9$

$$\frac{dy}{dx} = 3(\cdot 9)^2 - 6(\cdot 9) + 3$$

$$= 2\cdot 43 - 5\cdot 4 + 3$$

$= \cdot 03$ (i.e. $\frac{dy}{dx}$ is positive)

Let $x = 1\cdot 1$

$$\frac{dy}{dx} = 3(1\cdot 1)^2 - 6(1\cdot 1) + 3$$

$$= 3\cdot 63 - 6\cdot 6. + 3$$

$= \cdot 03$ (i.e. $\frac{dy}{dx}$ is positive)

Therefore at the point (1,3) the sign of $\frac{dy}{dx}$ is changing from positive through 0 to positive, and there is a point of inflexion.

Some points of inflexion can be identified, as in the above example, when $\frac{dy}{dx}$ is equal to 0, but is not changing sign. Others can only be identified by the use of the second derivative.

3. The Second Derivative

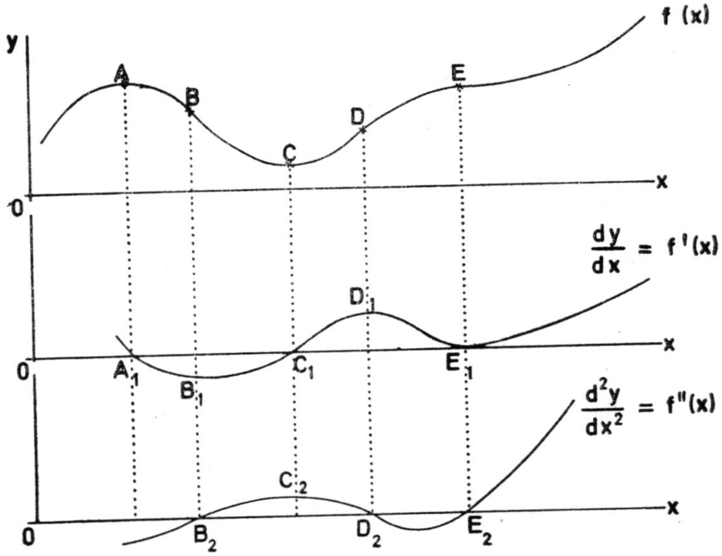

The first of the above sketches shows a curve $y = f(x)$ with
a maximum turning point at A
a minimum turning point at C
a point of inflexion at E.
(Points B and D will be considered shortly.)

The second sketch is the graph of $\frac{dy}{dx}$. Notice that $\frac{dy}{dx}$ is equal to 0 at points A_1, C_1, E_1; positive to the left of A_1, negative from A_1 to C_1, positive from C_1 to E_1, and to the right of E_1.

$\frac{dy}{dx}$ has a minimum turning point at B_1
a maximum turning point at D_1
a minimum turning point at E_1

The third sketch is the graph of the gradient of $\frac{dy}{dx}$ measured by the derivative of $\frac{dy}{dx}$ and known as the second derivative, $\frac{d^2y}{dx^2}$.

Notice that when $\frac{dy}{dx}$ has turning points, at B_1, D_1, and E_1, $\frac{d^2y}{dx^2}$ is equal to 0; therefore, at B_2, D_2 and E_2, $\frac{d^2y}{dx^2} = 0$.

When the gradient of $\frac{dy}{dx}$ is negative, to the left of B_1, and between D_1 and E_1, $\frac{d^2y}{dx^2}$ is negative; between B_1 and D_1 and after E_1, $\frac{d^2y}{dx^2}$ is positive.

Note

To calculate the second derivative, differentiate the first derivative.

Example

$$y = x^4 - 3x^3 + 4x^2$$

$$\frac{dy}{dx} = 4x^3 - 9x^2 + 8x$$

$$\frac{d^2y}{dx^2} = 12x^2 - 18x + 8$$

At A, the maximum turning point on the original curve,

$$\frac{dy}{dx} = 0, \quad \frac{d^2y}{dx^2} \text{ is negative.}$$

At C, the minimum turning point on the original curve,

$$\frac{dy}{dx} = 0, \quad \frac{d^2y}{dx^2} \text{ is positive.}$$

This result provides a new test for maximum and minimum turning points: to find turning points, let $\frac{dy}{dx} = 0$, and investigate the *sign* of the second derivative.

At B, D, and E $\frac{d^2y}{dx^2}$ is equal to 0, and each of these points on the original curve is known as a point of inflexion.

Points of inflexion occur when the rate of increase or decrease of the first derivative (as measured by $\frac{d^2y}{dx^2}$) changes. At B_1 and E_1, the gradient of the first derivative changes from negative to positive; at D_1, the gradient changes from positive to negative. In other words, B_1 and E_1 are minimum turning points on the graph of $\frac{dy}{dx}$ and D_1 is a maximum; the second derivative is equal to 0 and is changing sign at B_2, D_2 and E_2.

4. Summary of Results

For any curve $y = f(x)$

(i) If $\frac{dy}{dx} = 0$, there is a maximum or a minimum turning point, or a point of inflexion.

(ii) If $\frac{dy}{dx} = 0$, and $\frac{d^2y}{dx^2}$ is positive, there is a minimum turning point.

(iii) If $\frac{dy}{dx} = 0$, and $\frac{d^2y}{dx^2}$ is negative, there is a maximum turning point.

(iv) If $\frac{d^2y}{dx^2} = 0$, and is changing sign, there is a point of inflexion.

Examples

Find turning points and points of inflexion (if any) for the following curves:

(a) $y = x^2 + 12x - 20$

$$\frac{dy}{dx} = 2x + 12$$

$$\frac{d^2y}{dx^2} = 2$$

Let $\frac{dy}{dx} = 0$

$$2x + 12 = 0$$

$$x = -6$$

When $x = -6$, $\quad y = (6)^2 + 12(-6) - 20$
$= -56$

The sign of $\frac{d^2y}{dx^2}$ is positive,

Therefore there is a minimum turning point at $(-6, -56)$

Note No value of x will make the second derivative equal to 0, therefore there is no point of inflexion.

(b) $y = 40 + 3x - 2x^2 + \frac{x^3}{3}$

$$\frac{dy}{dx} = 3 - 4x + x^2$$

$$\frac{d^2y}{dx^2} = -4 + 2x$$

Let $\frac{dy}{dx} = 0.$

$$3 - 4x + x^2 = 0$$

$$(3 - x)(1 - x) = 0$$

$$x = 3 \text{ or } 1.$$

When $x = 3$

$$y = 40 + 3(3) - 2(3)^2 + \frac{(3)^3}{3}$$

$$= 40$$

$$\frac{d^2y}{dx^2} = -4 + 2(3)$$

$$= 2 \text{ (i.e. positive)}$$

Therefore there is a minimum at (3,40).

When $x = 2$

$$y = 40 + 3(1) + - 2(1)^2 + \frac{(1)^3}{3}$$

$$= 41\tfrac{1}{3}$$

$$\frac{d^2y}{dx^2} = -4 + 2(1)$$

$$= -2 \text{ (i.e. negative)}$$

Therefore there is a maximum at $(1, 41\tfrac{1}{3})$.

A point of inflexion will occur when $\frac{d^2y}{dx^2} = 0$, and is changing sign.

Let $\frac{d^2y}{dx^2} = 0$

$$-4 + 2x = 0$$

$$x = 2$$

When $x<2$, $\frac{d^2y}{dx^2}$ is negative

When $x>2$, $\frac{d^2y}{dx^2}$ is positive

Therefore a point of inflexion exists at $x=2$

$$y = 40 + 3(2) - 2(2)^2 + (\tfrac{2}{3})^3$$

$$= 40\tfrac{2}{3}$$

Answer

A minimum at (3,40)

A maximum at $(1, 41\tfrac{1}{3})$

A point of inflexion at $(2, 40\tfrac{2}{3})$.

EXERCISE 7.

Find maximum and minimum turning points and points of inflexion (if any) for the following curves. Give x and y coordinates of these points and make a rough sketch of the curve.

1. $y = x^2 - 10x + 12$
2. $y = 6 - 6x + x^2$
3. $y = 8 + 4x - x^2$
4. $y = x^3 - 27x + 3$

5. $y = x^3 - 6x^2 + 12x - 7$

6. $y = x^2 (x - 3)$

7. $y = x + \frac{1}{x}$

8. $y = \frac{x^3}{3} - 3x^2 + 9x - 13$

VIII. PARTIAL AND COMPLETE DIFFERENTIATION

1. Functions of More than One Variable

Many functions have more than one independent variable. A function with two independent variables could be expressed

$$y = f(x,z).$$

Similarly, a function with three independent variables could be expressed

$$y = f(w,x,z),$$

and in general a function with n independent variables may be written

$$y = f(x_1,\ x_2,\ x_3\ \ldots .\ x_n).$$

Note. Either different letters of the alphabet or subscripts may be used to denote different variables. In the latter case it is important to realise that x_1, x_2, x_3, etc. are all completely independent of one another.

2. Partial Differentiation

If a function has more than one independent variable the function may be differentiated with respect to any one of the variables if the others are assumed to remain constant. These results are known as *partial derivatives*.

To differentiate a function with respect to one of its variables, treat the remaining variables as constants and proceed in the usual way. To denote a partial derivative the Greek letter δ is used in place of the Roman d.

Examples

(a) Give the partial derivatives of

$$y = x^2 + 3xz - 4z^2$$

$$\frac{\delta y}{\delta x} = 2x + 3z$$

$$\frac{\delta y}{\delta z} = 3x - 8z$$

(b) Give the partial derivatives of

$$y = w^3 - w^2x + x^2z - z^2.$$

$$\frac{\delta y}{\delta w} = 3w^2 - 2wx.$$

$$\frac{\delta y}{\delta x} = - w^2 + 2xz$$

$$\frac{\delta y}{\delta z} = x^2 - 2z.$$

3. Complete Differentiation

(i) *Functions with Two Independent Variables*

If a function has two independent variables

$$y = f(x,z)$$

then y may change

1. If x changes but z remains constant.
2. If z changes but x remains constant.
3. If both x and z are changing.

In the first case the rate of change may be measured by the partial derivative $\frac{\delta y}{\delta x}$ and for small changes in x the corresponding change in y may be approximately calculated by the use of the expression

$$\triangle y \approx \frac{\delta y}{\delta x}(\triangle x) \quad \text{(See VI D)}$$

In the second case the rate of change may be measured by the partial derivative $\frac{\delta y}{\delta z}$ and for small changes in z

$$\triangle y \approx \frac{\delta y}{\delta z}(\triangle z).$$

In the third case, when x and z are changing simultaneously

$$\triangle y \approx \frac{\delta y}{\delta x}(\triangle x) + \frac{\delta y}{\delta z}(\triangle z).$$

This expression, which shows how small changes in x and z ($\triangle x$ and $\triangle z$) lead to a change in the value of y ($\triangle y$) is known as the *complete differential* of $y = f(x,z)$.

Note. The complete differential is normally given as

$$dy = \frac{\delta y}{\delta x}(dx) + \frac{\delta y}{\delta z}(dz)$$

The approximate form for $\triangle y$ is used here to emphasize the inexactitude of any numerical calculation using the complete differential.

Examples

(a) Give the complete differential of

$$y = 4x^2 - 5xz + \frac{1}{z}$$

$$\frac{\delta y}{\delta x} = 8x - 5z$$

$$\frac{\delta y}{\delta z} = -5x - \frac{1}{z^2}$$

Therefore

$$\Delta y \approx (8x - 5z)(\Delta x) + (-5x - \frac{1}{z^2})(\Delta z).$$

(b) If $y = x^2 + 2xz - z^2$,
calculate the approximate change in y if x increases from 2 to 2·1 and z increases from 3 to 3·2

$$\frac{\delta y}{\delta x} = 2x + 2z$$

$$\frac{\delta y}{\delta z} = 2x - 2z$$

$$\Delta x = \cdot 1$$

$$\Delta z = \cdot 2$$

Therefore the complete differential

$$\Delta y \approx \frac{\delta y}{\delta x}(\Delta x) + \frac{\delta y}{\delta z}(\Delta z)$$

$$\approx (2x + 2z)(\cdot 1) + (2x - 2z)(\cdot 2)$$

Substituting 2 for x and 3 for z

$$\Delta y \approx (4 + 6)(\cdot 1) + (4 - 6)(\cdot 2)$$

$$\approx (10)(\cdot 1) + (-2)(\cdot 2)$$

$$\approx 1 - \cdot 4$$

$$\approx \cdot 6$$

Answer y increases by approximately ·6.

[This solution may be checked by numerical calculation.

When $x = 2, z = 3$

$$y = (2)^2 + 2(2)(3) - (3)^2$$

$$= 7$$

When $x = 2\cdot 1, z = 3\cdot 2$

$$y = (2\cdot 1)^2 + 2(2\cdot 1)(3\cdot 2) - (3\cdot 2)^2$$

$$= 7\cdot 61$$

Therefore the actual increase in y was ·61]

(ii) *Functions with Three or More Independent Variables*
If a function has three or more independent variables, the complete differential may be calculated in the same way.

Thus for a three-variable function $y = f(w,x,z)$

$$\Delta y \approx \frac{\delta y}{\delta w}(\Delta w) + \frac{\delta y}{\delta x}(\Delta x) + \frac{\delta y}{\delta z}(\Delta z).$$

(where Δw, Δx and Δz represent *small* changes in w, x and z).
In general for the function

$$y = f(x_1, x_2, x_3, \ldots\ldots\ldots x_n)$$

the effect of small changes in x_1, x_2, x_3, etc. may be measured by the complete differential

$$\Delta y \approx \frac{\delta y}{\delta x_1}(\Delta x_1) + \frac{\delta y}{\delta x_2}(\Delta x_2) \ldots + \frac{\delta y}{\delta x_n}\Delta x_n)$$

Example
If $y = 4w^2 - 3wx^2 + xz + 4z^2$
calculate the approximate change in y if:
w increases from 2 to 2·1
x increases from 3 to 3·2
z decreases from 2.5 to 2·4

$$\frac{\delta y}{\delta w} = 8w - 3x^2 \qquad \Delta w = \cdot 1$$

$$\frac{\delta y}{\delta x} = -6wx + z \qquad \Delta x = \cdot 2$$

$$\frac{\delta y}{\delta z} = x + 8z \qquad \Delta z = -\cdot 1$$

$$\Delta y \approx \frac{\delta y}{\delta w}(\Delta w) + \frac{\delta y}{\delta x}(\Delta x) + \frac{\delta y}{\delta z}(\Delta z)$$

$$\approx (8w - 3x^2)(.1) + (-6wx + z)(\cdot 2) + (x + 8z)(-\cdot 1)$$

Substituting : $w = 2$, $x = 3$, $z = 2\cdot 5$

$$\Delta y \approx (16 - 27)(\cdot 1) + (-36 + 2\cdot 5)(\cdot 2) + (3 + 20)(-\cdot 1)$$

$$\approx -1\cdot 1 \text{ x } - 6.7 - 2\cdot 3$$

$$\approx -10\cdot 1$$

Answer y decreases by approximately 10.1

EXERCISE 8.

1. Give the partial derivatives of the following functions with respect to x and with respect to z:

   (a) $y = x^2 - 4xz + z^2$

   (b) $y = x^2z + 4xz - 3xz^2$

   (c) $y = 6xz + 4x - \frac{1}{z}$

   (d) $y = 2x^{\frac{1}{2}} - xz - 5z$

   (e) $y = (x^2 - z^3)^7$

2. $y = 2x^2 + 5xz + 3z^2$

   (a) Give the complete differential of the above function.

   (b) Using the result of (a), calculate the approximate increase in y when x increases from 4 to 4·1 and z increases from 5 to 5·1.

   (c) Check the result of (b) by numerical calculation

      (i) How large an error results from using the complete differential?

      (ii) Express this error as a per cent of the total increase in y.

3. $y = 3x^2 - 4xz - z^2$

   (a) Give the complete differential of the above function.

   (b) Show that if x increases from 3 to 3.1, and z increases from 1 to 1.2, the value of y decreases by approximately 1.4.

4. $y = 3w^2 + 2wx^2 - 5xz^2 + z$

   (a) Give the complete differential of the above function.

   (b) Calculate the approximate change in y if:

   w increases from 1 to 1·1
   x decreases from 3 to 2·9
   z increases from 4 to 4·2.

IX. INTEGRATION

1. Introduction

The concept of an *integral* may be approached in two ways. In the first place, an integral is the result of reversing the process of differentiation. Given a derivative, the process of integration enables us to find the original function. This kind of integral is known as an *indefinite integral.* Secondly, an integral may be regarded as the limit of a certain addition process and may be graphically interpreted as the area enclosed by a curve or a set of curves. This kind of integral is known as a *definite integral.*

2. The Indefinite Integral

What function when differentiated gives

$$\frac{dy}{dx} = 3x^2 \ ?$$

Inspection shows that the function

$$y = x^3$$

has this derivative. However, the functions

$$y = x^3 + 2$$
$$y = x^3 - 5$$
$$y = x^3 - 200 \text{ etc.}$$

also have the derivative

$$\frac{dy}{dx} = 3x^2$$

so it is necessary to say that the function whose derivative is $3x^2$ is

$$y = x^3 + C,$$

where C is any arbitrary constant.

This process of finding the original function when the derivative is known is called integration, and the original function is known as the integral. The integral sign " $\int$ " is always accompanied by the expression "dx" if the expression to be integrated is a function of x. Thus the above problem is written

$$\int(3x^2)dx = x^3 + C.$$

For simple functions of x, the following general rule may be given:

$$\int (ax^n)dx = \frac{ax^{n+1}}{n+1} + C.$$

Examples

(a) $\int (x^5)dx = \frac{x^6}{6} + C$

(b) $\int (6x^2)\,dx = 2x^3 + C$

c) $\int (2x^3 + 3x^2 - 4x)\,dx = \frac{1}{2}x^4 + x^3 - 2x^2 + C.$

It is often difficult to find the integrals of more complicated functions, and few rules can be given. Sometimes, an expression is obviously the derivative of some function, and in this case the integral can be found.

Example

$\int 40x\,(5x^2 - 3)^3\,dx = (5x^2 - 3)^4 + C.$

3. The Definite Integral

(i) *A Geometrical Problem*

In the same way that the concept of a derivative was introduced as the answer to a problem in co-ordinate geometry (how to measure the slope of a curve at a point) so the concept of an integral may be introduced as an answer to a geometrical problem. Given the curve y= f(x), how can we calculate the area under the curve and above the x-axis between the points A and B?

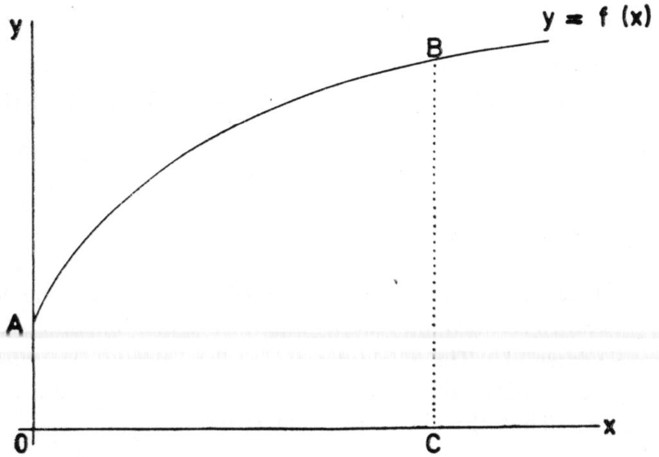

To calculate: Area OABC

While no method of calculating the area OABC is yet known, the calculation of the area of a rectangle presents no difficulty. We therefore introduce AD and EB parallel to the x-axis.

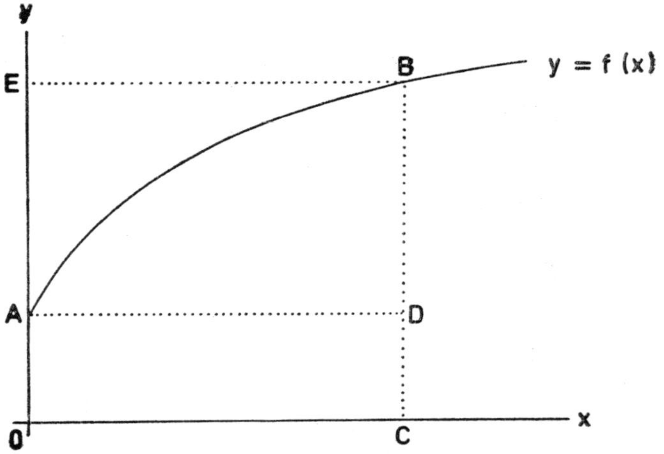

We can safely say that the wanted area OABC lies between the area of the large rectangle OEBC, and the small rectangle OADC. In other words

$$\text{OEBC} > \text{OABC} > \text{OADC}.$$

This does not give a very exact solution to the problem, but greater precision may be given by taking a point F on the curve $y = f(x)$ between A and B. The rectangle analysis may be repeated for the area under the curve to the left of F, and to the right of F.

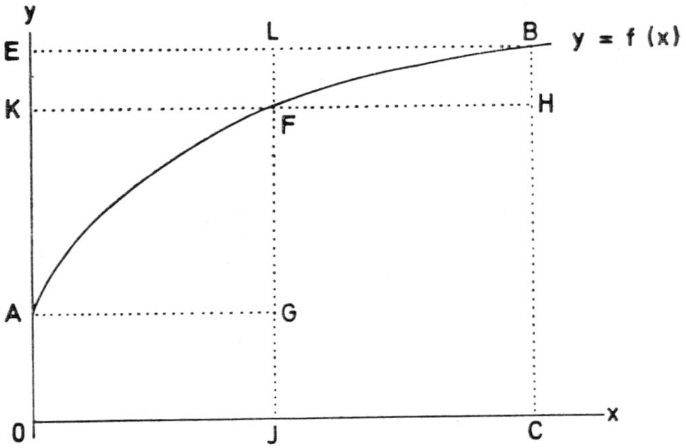

The wanted area OABC is now divided into two parts OAFJ and JFBC. Of the left hand part we can say that the area of OAFJ lies between the area of the large rectangle OKFJ and the area of the small rectangle OAGJ. Similarly, area JFBC lies between the area of the large rectangle JLBC and the area of the small rectangle JFHC. Combining these results, the wanted area OABC lies between the combined areas of the two large rectangles, and the combined areas of the two small rectangles:

OKFJ + JLBC > OABC > OAGJ + JFHC.

This description of the area OABC is more precise than our earlier one.

If many points are chosen between A and B, the area OABC can be described in terms of the areas of the many strips under AB.

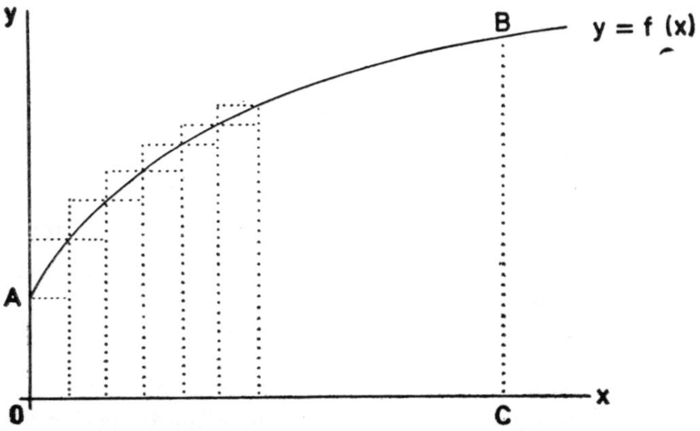

If the strips are close enough together, the sum of the areas of the large rectangles will be very little larger than the wanted area and the sum of the areas of the small rectangles will be very little smaller.

(ii) *An Algebraic Solution*

With this analysis in mind, consider the curve y = f(x), with two points P (x,y) and Q (x + Δx, y + Δy):

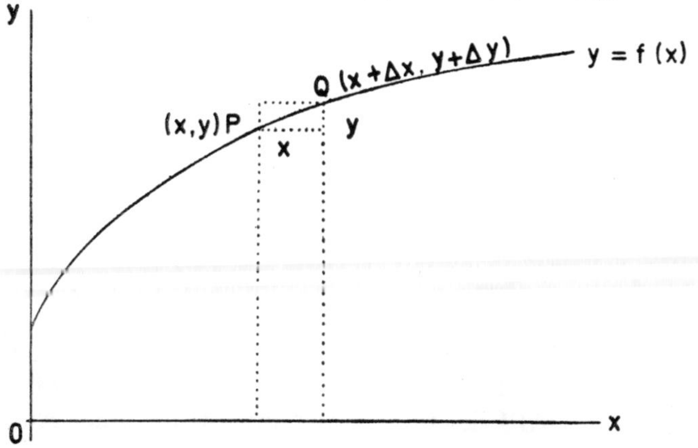

The area under the curve between P and Q may be thought of as one of the strips making up the total area (A) under the curve $y = f(x)$. This strip of area may be referred to as $\triangle A$. Inspection shows that the area of the large rectangle will be $(y + \triangle y)(\triangle x)$ and the area of the small rectangle will be $(y)(\triangle x)$. Hence

$$(y + \triangle y)(\triangle x) > \triangle A > (y)(\triangle x)$$

Dividing through by $\triangle x$

$$y + \triangle y > \frac{\triangle A}{\triangle x} > y$$

As $Q \rightarrow P$

$$\triangle x, \triangle y \rightarrow 0$$

$$\frac{\triangle A}{\triangle x} \rightarrow \frac{dA}{dx}$$

and the expression becomes

$$y > \frac{dA}{dx} > y.$$

In other words

$$\frac{dA}{dx} \rightarrow y.$$

$\frac{dA}{dx}$ is a derivative. It is the derivative of a function

$$A = f(x).$$

Therefore, if we can find a function whose derivative $\frac{dA}{dx}$ is equal to y, we can find the area A.

(iii) *A Numerical Example*

Given the curve $y = 2x$, what is the area under the curve between $x = 0$, and $x = 3$?

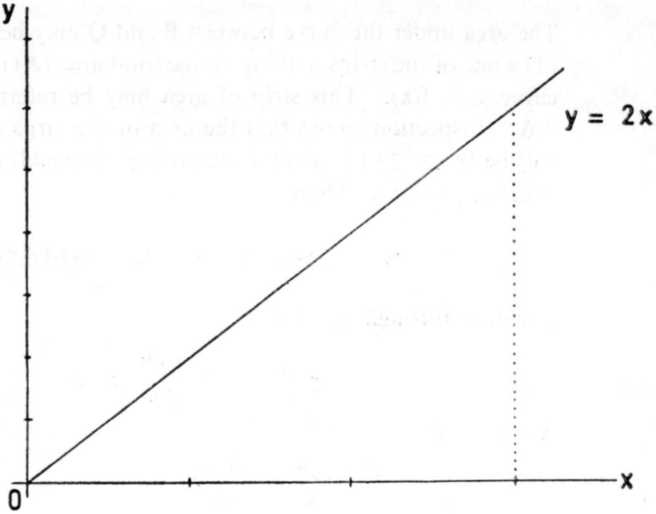

The area (A) under the curve is given by that expression whose derivative $\frac{dA}{dx}$ is equal to y.

$$y = 2x,$$

therefore we are looking for

$$A = \int_0^3 (2x)dx.$$

This integral, which shows the limiting values of x (x = 0 and x = 3) is a definite integral, and the *rule for the evaluation of a definite integral* is given by

$$\int_a^b f(x)\ dx = F(b) - F(a)$$

where a and b are numbers and F is that function whose derivative is f.

In the above case

$$A = \int_0^3 (2x)dx = \Big[x^2\Big]_0^3 = \Big[(3^2) - (0^2)\Big] = 9$$

The area under the curve y = 2x between the origin and x = 3 is 9 square units.

Note
The constant term C is ignored when calculating the definite integral since it would disappear when F(a) is subtracted from F(b).

(iv) *Further Examples*

(a) Find the area under the curve $y = x^3$ between the origin and x = 4.

$$A = \int_0^4 (x^3)\,dx = \left[\frac{x^4}{4}\right]_0^4$$

$$= \left[\frac{4^4}{4} - \frac{0^4}{4}\right]$$

$$= \frac{256}{4} = 64 \text{ square units.}$$

(b) Find the area under the curve $y = 3x^2$ between $x = 2$ and $x = 5$.

$$A = \int_2^5 (3x^2)dx = \left[x^3\right]_2^5$$

$$= 5^3 - 2^3$$

$$= 125 - 8$$

$$= 117 \text{ square units.}$$

EXERCISE 9

A. Evaluate:

1. $\int(x^3)dx.$

2. $\int(3x^2)dx.$

3. $\int(4x^3 + 2x^2 - 1)dx.$

4. $\int(9x^2 - 4x - 6)dx.$

5. $\int(3x)(2x^2 + 5x - 4)\,dx.$

6. $\int(x^2 + 2 + \frac{1}{x^2})dx.$

B. Evaluate:

1. $\int_0^3 (x^5)\,dx.$

2. $\int_2^4 (3x^2 + 2)\,dx.$

3. $\int_3^6 (x - 3)\,dx$

4. $\int_{1}^{2} \frac{(x^3 - 1)}{x^2} dx.$

C. 1. Find the area under the curve $y = 4x$ between the origin and $x = 5$.

2. Find the area under the curve $y = \frac{x^3}{3}$ between the origin and $x = 2$.

3. Find the area enclosed by the curve $y = \frac{x^3}{2}$, the x-axis, the y-axis and $x = 3$.

4. Find the area enclosed by the curve $y = 3x^5$, the x-axis, the y-axis and $x = \frac{1}{2}$.

D. 1. Find the area under the curve $y = x^4$ between $x = 1$ and $x = 2$.

2. Find the area under the curve $y = x^2$ between $x = 2$, and $x = 4$.

3. Find the area beneath the curve $y = x^5$, between $x = 2$ and $x = 3$,

4. Find the area beneath the curve $y = \sqrt{x}$ between $x = 4$, and $x = 9$.

E. 1. (a) Differentiate $y = (2x + 5)^3$

(b) Integrate $\frac{dy}{dx} = 6(2x + 5)^2$

(c) Hence find a general expression for the integral of $(ax + b)^n$.

2. The gradient of a curve at any point is given by

$$\frac{dy}{dx} = 2x - 1.$$

If the curve passes through the point (1,1), find the equation of the curve.

(Hint: when integrating do not forget the constant, C.)

3. Find the equation of the curve which passes through the point (0,1) and has the gradient $(1 - 2x^2)$.

X. CALCULUS AND THE THEORY OF THE FIRM

1. Introduction

We have already seen that almost all branches of economic theory are concerned with functional relationships. For example, it is usual to assume that the quantity of a good demanded depends upon its price, or, in mathematical terminology

$$Q = f(P)$$

In national income theory it is assumed that the level of expenditure on consumption goods depends upon the level of disposable income.

$$C = f(Y)$$

In the theory of the firm several important functional relationships are distinguished; on the production side the level of output is determined by the quantity of the factor inputs:

$$O = f(F_1, \ F_2, \ F_3, \ldots . F_n)$$

Costs are related to the quantity of goods produced:

$$C = f(Q)$$

Revenue, too, is determined by the quantity of goods sold

$$R = f(Q).$$

Differentiation was introduced as the process of measuring the change in the dependent variable caused by small changes in the independent variable. In the preceding Chapters we dealt with functions of the form

$$y = f(x);$$

the derivative which measured the limit of $\triangle y/\triangle x$ as $\triangle x \rightarrow 0$ was known as $\frac{dy}{dx}$.

The functions of economic theory can also be differentiated: the derivatives or partial derivatives of the functions mentioned above would be

$$\frac{dQ}{dp}, \quad \frac{dC}{dY}, \quad \frac{\delta O}{\delta F}, \quad \frac{dC}{dQ}, \quad \text{and} \quad \frac{dR}{dQ}.$$

We now have to see what economic meaning can be given to these derivatives. The following illustrations will be drawn mainly from the theory of the firm; but clearly calculus has a much wider application. Whenever it makes sense to express an economic relationship in functional form, the derivative can be used to measure the rate of change of the function as the independent variable(s) alter in value.

2. Derivatives and the Marginal Concept

(i) *A Production Function*

Let us consider the case of a farmer who wishes to analyse the effect of a new fertilizer on a crop of maize. He experiments, and finds that the output of maize (O) is affected by the number of loads of fertilizer (F) employed

$$O = f(F).$$

Furthermore, he realizes that the law of diminishing returns applies; there is a limit to the amount of fertilizer he can usefully apply. As he increases the quantity of fertilizer per acre, the extra load seems to make a smaller contribution to total output and more than twenty loads actually reduces the quantity of maize that he can grow. The graph relating output to fertilizer looks like this

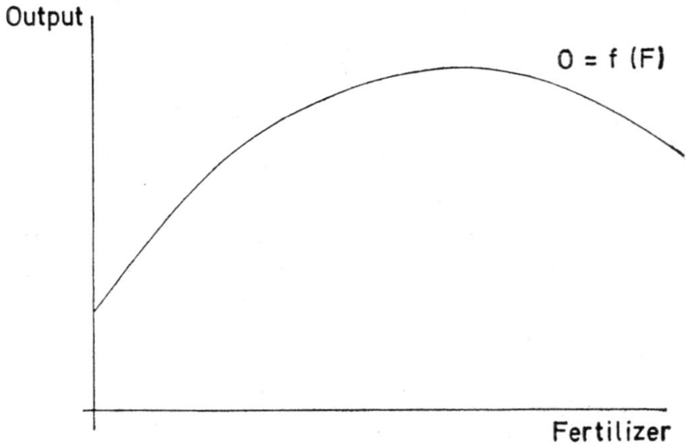

and could suitably be described by the function

$$O = 200 + 40F - F^2.$$

The derivative of the function is

$$\frac{dO}{dF}, \text{ the limit of } \frac{\triangle O}{\triangle F} \text{ as } \triangle F \rightarrow 0.$$

$\frac{\triangle O}{\triangle F}$ measures the rate of increase in output as fertilizer is increased, or the increase in output following from the employment of an additional unit of fertilizer. This is a familiar economic concept known as the *marginal product* of the factor input — in this case, fertilizer.

We know that the derivative measures the *slope* of a curve. The marginal product $\frac{\Delta O}{\Delta F}$ is represented by the slope of the total product curve.

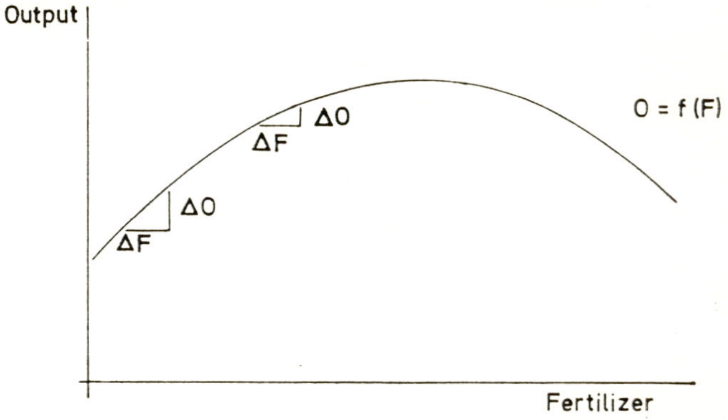

It is clear that $\frac{\Delta O}{\Delta F}$, the slope of the curve, gradually decreases in value; showing that equal additions of fertilizer do not yield equal increments of maize output. Indeed, as the farmer discovered, after 20 loads of fertilizer the extra load actually diminishes total output; $\frac{\Delta O}{\Delta F}$ is negative.

Note. It should be understood that when we talk of the derivative of this function we mean $\frac{dO}{dF}$, the limit of $\frac{\Delta O}{\Delta F}$ as $\Delta F \rightarrow 0$.

In practical terms this is rather an unreal concept; either the farmer uses an extra load of fertilizer or he does not. Hence the derivative which measures the slope at a point provides only an approximate measure for the marginal product. For most purposes, however, the approximation is near enough for us to say that the derivative of the total product curve measures the marginal product.

In the same way that the derivative of a production function $O = f(F)$ measures the marginal product of F, so the derivative of a cost function $C = f(Q)$, $\frac{dC}{dQ}$, measures the cost of the extra quantity produced, usually known as the *marginal cost*. Again, the *slope* of the cost curve, measured by the derivative, shows the approximate marginal cost at any level of output.

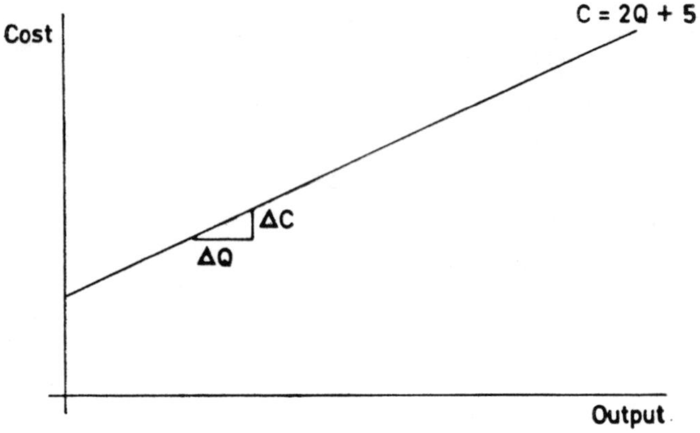

In this case the derivative of the function is 2; the function is linear so that the slope is constant throughout; the economic implication is that for this particular firm marginal cost is the same, i.e. 2, at all levels of output.

Similarly any total revenue function R=f(Q) will have a derivative $\frac{dR}{dQ}$ which measures the way in which revenue (R) varies with the quantity sold (Q). At any point the slope of the total revenue curve will measure *marginal revenue.*

(ii) *The Use of Partial and Complete Derivatives*

The use of partial derivatives may also be illustrated by means of a production function. Let us assume that the output of a firm is determined by the amount of capital invested in machinery (one unit of K = £1000) and by the number of men employed (one unit of labour L = 10 men). In this case

$$O = f(K,L)$$

and we will assume that the actual equation is

$$O = 20KL - 3K^2 - 4L^2$$

The two partial derivatives of this function measure the marginal products of the two factors of production. Marginal product of capital (K)

$$= \frac{\delta O}{\delta K} = 20L - 6K$$

Marginal product of labour (L)

$$= \frac{\delta O}{\delta L} = 20K - 8L.$$

If £7000 of capital and 90 men are employed, $K = 7$, $L = 9$, and we have

marginal product of capital	=	20(9) − 6(7)
	=	138
marginal product of labour	=	20(7) − 8(9)
	=	68

It would appear that this firm could raise its output by using more capital and less labour. (Why?)

Suppose the firm decides to sack 2 men and invest another £100 in capital equipment. L will have decreased by ·2; K will have increased by ·1. We have enough information to calculate the change in output resulting from this decision. The *complete differential* measures the change in the dependent variable (O) as the two independent variables (K and L) change. The original function was

$$O = 20KL - 3K^2 - 4L^2.$$

The complete differential is

$$\Delta O \approx (20L - 6K)(\Delta K) + (20K - 8L)(\Delta L)$$

We know $K = 7$, $L = 9$, $\Delta K = \cdot 1$, $\Delta L = -\cdot 2$.

$$\Delta O \approx [20(9) - 6(7)][\cdot 1] + [20(7) - 8(9)][-.2]$$

$$\approx 13{\cdot}8 - 13.6$$

$$\approx 0{\cdot}2$$

Output has risen slightly as a result of this decision to substitute machines for men.

Notice that the actual decision to employ more machinery and fewer men will depend upon the price of the factors and the price of the product. If the monthly wage is £20, the firm will have saved £40 by sacking the two employees, but the additional capital cost £100. The additional outlay of £60 will only have been worthwhile if the extra ·2 units of output sells for more than £60.

3. The "Average" Concept in Economics

In the preceding paragraph the concept of the *marginal* item was discussed, and it was shown that the derivative of a function measures the approximate magnitude of the marginal item:

the derivative of a production function measures the marginal product, the derivative of a total cost function measures marginal cost, and the derivative of a total revenue function measures marginal revenue. In each of these cases the derivative measures the slope of the graph of the function at any point.

Another concept frequently used in economic theory is that of the *average*; for example, we speak of the average output per worker, the average cost of each item produced, and the average revenue per item sold. The word "average" is used in economics in exactly the same way as it is in ordinary speech. If we wish to know the average height of a group of five people, we first find the total height of the group by adding the five individual heights together, and then divide by the number of people in the group to find the average height per person. Similarly, if we have an expression showing how total output varies with the numbers of workers employed, $O = f(L)$, then the average output per worker is equal to total output divided by the number of workers employed.

Example

Suppose the following expression shows how output on a farm (O) varies with the number of men employed (L):

$$O = 12L - L^2.$$

Find an algebraic expression for the average output per worker and hence calculate the average output per worker when 3 men are employed on the farm.

$$\text{Average output} = \frac{O}{L} = \frac{12L - L^2}{L} = 12 - L.$$

When three men are employed $L = 3$ and the average output per worker $= 12 - L = 12 - 3 = 9$.

(Check: when $L = 3$, total output $= 12L - L^2 = 36 - 9 = 27$.

Average output is equal to total output divided by the number of men employed $= \frac{27}{3} = 9$).

If a total cost function is of the form $C = f_1(Q)$ where Q represents the number of items produced, then average cost will be equal to total cost divided by the number of items produced:

$$C = f_1(Q); \; AC = \frac{f_1(Q)}{Q}$$

Average revenue is related to total revenue in the same way:

$$R = f_2(Q); \; AR = \frac{f_2(Q)}{Q}.$$

This last case is one of the most interesting to the economist because the average revenue curve is the "demand curve" of the firm. (From the point of view of the producer "average revenue" is the average amount he can charge for each item sold to the public; from the point of view of the consumer this amount charged is of course the price.) Average revenue (and therefore price) will under certain market conditions be different for different levels of output offered for sale. If the firm produces for a perfectly competitive market, the demand curve will be horizontal; if the firm sells in a monopolistic market the demand curve will be downward sloping. This can be illustrated mathematically. Let us suppose there are two firms A and B with the following revenue functions;

Firm A	*Firm B*
$R = 23Q$	$R = 40Q - Q^2$
$AR = \frac{23Q}{Q} = 23$	$AR = \frac{40Q - Q^2}{Q} = 40 - Q$

The graphs of these two average revenue functions are:

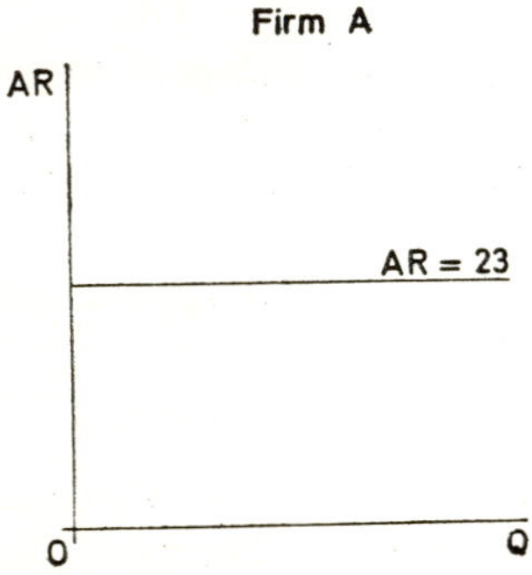

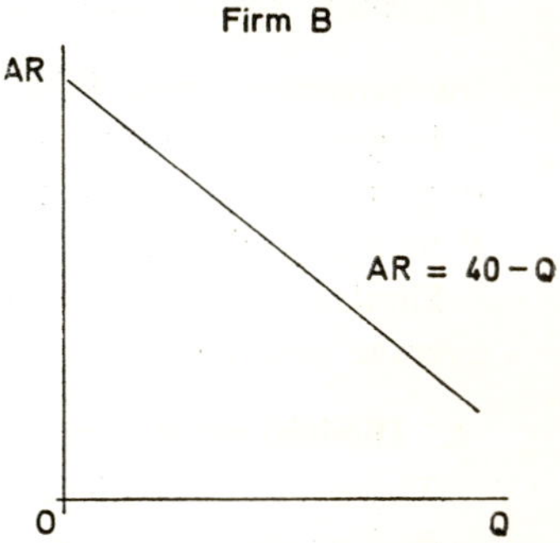

Clearly Firm A produces for a perfectly competitive market; Firm B has some degree of monopoly. (An interesting exercise would be to find the derivative of the total revenue curves which would give the marginal revenue functions; plot them on the above graphs and compare the result with diagrams showing competitive and monopolistic markets in any textbook of economics.)

A method of calculating averages from a graph or diagram may be illustrated here. The following diagram represents a production function showing how the output of maize (M) varies with the number of men employed (L):

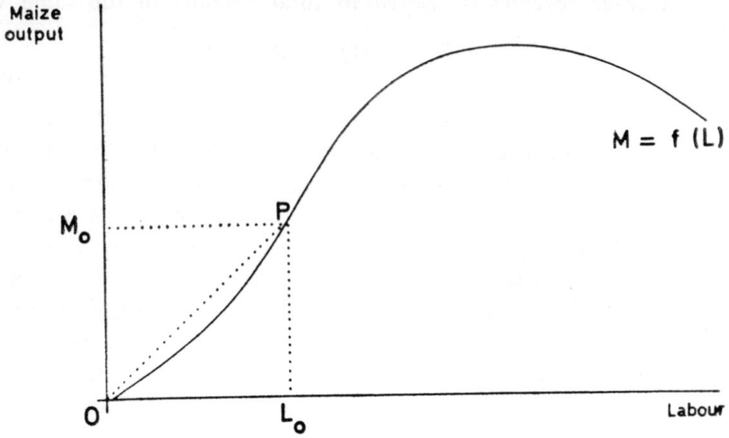

At point P the total output is M_O and the number of people employed is L_O. The average output per worker at this point is M_O divided by L_O or $\frac{OM_O}{OL_O}$. Geometrically speaking, the height OM_O is equivalent to the height PL_O so the average output could be written

$$\text{average output} = \frac{OM_O}{OL_O} = \frac{PL_O}{OL_O}.$$

Inspection of the diagram shows that the ratio $\frac{PL_O}{OL_O}$ measures the *tangent* of angle POL_O and therefore the slope of the line OP.

In general terms the average output at any point P on the total output curve is shown by the slope of the line joining the point P to the origin O.

Similar arguments apply to the graphical representation of average cost and average revenue.

4. Diminishing Returns

There are two distinct concepts in economics which are described by the name "diminishing returns". The more usual one may be defined as follows:

> "The law of diminishing returns refers to the amount of extra output that we get when we successively add equal extra units of a varying input to a fixed amount of some other input." (P. Samuelson, *Economics.*)

Let us look at the mathematical interpretation of this statement. Suppose that output (O) is a function of the three traditional inputs (or factors of production) land (N), labour (L) and capital (K)

$$O = f(N,L,K)$$

We are interested in the effect on output when one input is varied and the other two are held constant. The "extra output" from the "extra units of a varying input" may be measured by the ratios $\frac{\triangle O}{\triangle N}, \frac{\triangle O}{\triangle L}, \frac{\triangle O}{\triangle K}$ or (approximately) by the limiting values of these ratios given by the partial derivatives $\frac{\delta O}{\delta N}, \frac{\delta O}{\delta L}, \frac{\delta O}{\delta K}$.

> "An increase in some inputs relative to other fixed inputs (continues Samuelson) will cause total output to increase; but after a point the extra output resulting from the same additions of extra inputs is likely to become less and less".

In other words the traditional law of diminishing returns states that the marginal product of any factor input will decline or "diminish" after a certain point.

The other definition of diminishing returns relates to the size of the marginal product to the size of the average product at any point.

We are interested in the exact relationship between the extra unit of input and the resulting extra unit of output. For example, if there is a 3% increase in the factor input does output increase by less than 3%, by exactly 3%, or by more than 3%? If we let $\frac{\triangle O}{O}$ denote the proportionate increase in output and $\frac{\triangle K}{K}$ denote the proportionate increase in the factor input there are three possibilities:

$$\frac{\triangle O}{O} < \frac{\triangle K}{K} \quad \text{Diminishing returns}$$

$$\frac{\triangle O}{O} = \frac{\triangle K}{K} \quad \text{Constant returns}$$

$$\frac{\triangle O}{O} > \frac{\triangle K}{K} \quad \text{Increasing returns}$$

Using the rules for multiplying inequalities, these may also be written

$$\frac{\triangle O}{\triangle K} < \frac{O}{K} \quad \text{Diminishing returns}$$

$$\frac{\triangle O}{\triangle K} = \frac{O}{K} \quad \text{Constant returns}$$

$$\frac{\Delta O}{\Delta K} > \frac{O}{K} \quad \text{Increasing returns.}$$

We know that $\frac{\Delta O}{\Delta K}$ represents the marginal product of capital; $\frac{O}{K}$ represents the average product. Hence expressing the relationships mathematically enables us to say:

"At any point on a production function (or total output curve) diminishing returns exist if the marginal product is smaller than the average product; constant returns exist if the marginal product is equal to the average product; increasing returns exist if the marginal product is greater than the average product."

Example

Suppose the production function relating agricultural output to varying amounts of capital input is of the following form:

$$O = \frac{-K^3}{3} + 2K^2 + 12K.$$

Beyond what point do diminishing returns exist?

We can approach this problem in two ways: graphically and algebraically.

The graph of the function looks like this:

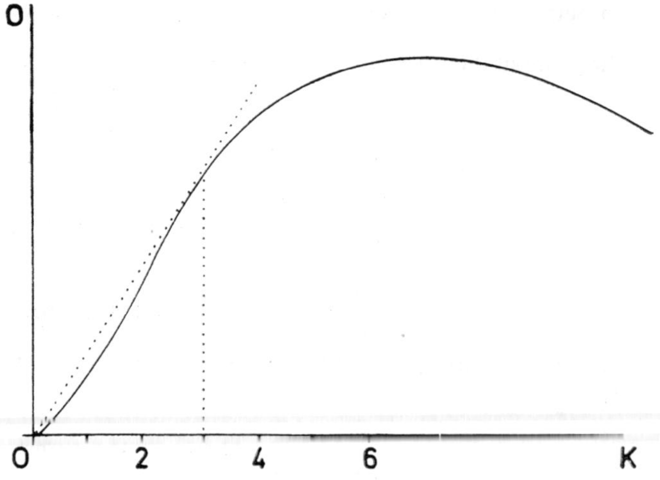

The average product (measured by the slope of the line from the origin to the curve) is equal to the marginal product

(the slope of the curve) at about the point where $K = 3$. Before this point marginal product is clearly greater than average product; after this point it is less.
By algebra

$$O = \frac{-K^3}{3} + 2K^2 + 12K$$

$$MP = \frac{\Delta O}{\Delta K} = -K^2 + 4K + 12$$

$$AP = \frac{O}{K} = \frac{-K^2}{3} + 2K + 12$$

MP = AP when

$$-K^2 + 4K + 12 = \frac{-K^2}{3} + 2K + 12$$

$$-K^2 + \frac{K^2}{3} + 4K + 2K = O$$

$$-2K^2 + 6K = O$$

$$K(-2K + 6) = O$$

either $K = 0$, *or* $K = 3$.

The first answer is uninteresting from an economic point of view, therefore $K = 3$ is the solution to the problem; three is the value of K for which the marginal product is equal to the average product. Anywhere beyond this point we can see from the graph the marginal product is smaller than the average product and diminishing returns set in.

5. Maximum and Minimum Problems

Many economic and business problems are concerned with finding maximum or minimum values. Maximum output, minimum costs and maximum profits are familiar concepts to any businessman. We shall consider two problems here: first a problem of minimizing costs, second a problem of profit maximization. (The student who wishes to try a problem of output maximization should calculate the maximum turning point for the production function in the above example and ask himself why in economic terms output is at a maximum when the marginal product [the derivative] is equal to zero).

(i) *Minimizing Costs*

A typical cost function will show the fixed costs of the business (the rent, fixed interest payments, etc. that must be paid even if there is no output) and will show how costs increase with increasing output. For example, suppose

$$C = \frac{1}{10}Q^2 + 5Q + 200$$

where C = total cost, and Q = quantity produced. In this function the fixed costs are 200; mathematically, when Q = 0, C = 200. Here is a rough sketch of the relevant section of the curve.

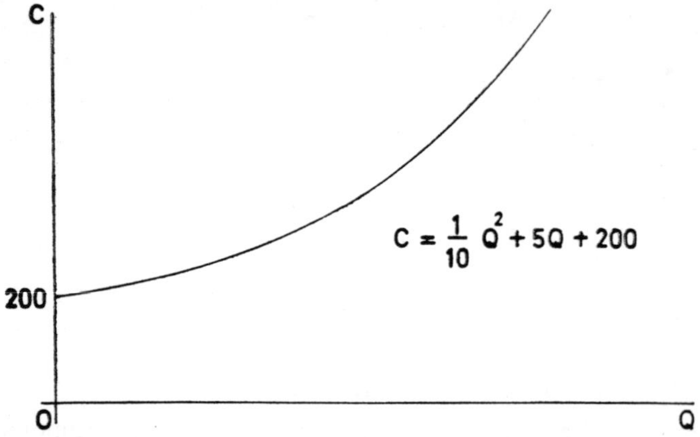

Clearly total cost increases steadily as output increases. This is what one would expect: it would be an odd case where *total* costs fell as output increases. When we speak of minimizing costs therefore, we do not mean minimizing total cost, but minimizing the *average* costs of production. For the above function

$$AC = \frac{1}{10} Q + 5 + \frac{200}{Q}$$

This function is at a minimum when its derivative is equal to zero.

$$\frac{dAC}{dQ} = \frac{1}{10} - \frac{200}{Q^2}$$

$$\text{Let} \quad \frac{dAC}{dQ} = \frac{1}{10} - \frac{200}{Q^2} = 0$$

$$\frac{1}{10} = \frac{200}{Q^2}$$

$$Q^2 = 2000$$

$$Q = \pm 44{\cdot}72$$

We ignore the negative answer, and decide that Average Cost is at a minimum when Q = 44·72. (Note: the second derivative is positive, $\frac{400}{Q^3}$, therefore this is a minimum and not a maximum turning point.)

Common sense tells us that Average Cost will be at a minimum at the point where it is equal to Marginal Cost. If the marginal cost is less than the average cost each additional item produced will lower the average; the average cost curve will be falling. If marginal costs are greater than the average cost, each additional item will raise the average; the average cost curve will be rising. Therefore the marginal cost curve will always cut the average cost curve at its lowest point.

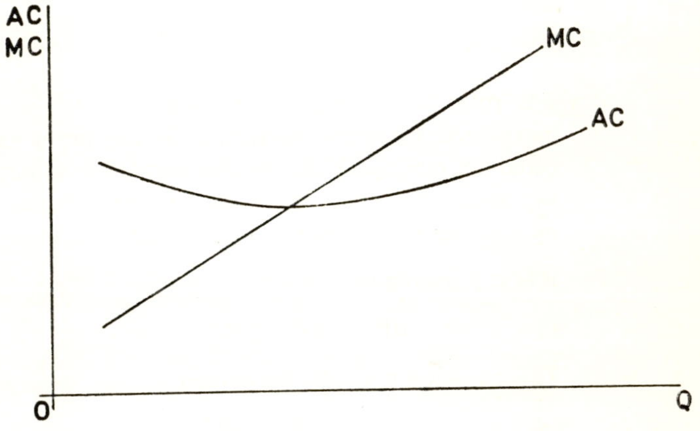

Hence there is another method of finding the point of lowest average cost: let average cost be equal to marginal cost and solve for Q. Using the above cost function:

$$C = \frac{1}{10}Q^2 + 5Q + 200$$

$$AC = \frac{1}{10}Q + 5 + \frac{200}{Q}$$

$$MC = \frac{1}{5}Q + 5$$

Let AC = MC

$$\frac{1}{10}Q + 5 + \frac{200}{Q} = \frac{1}{5}Q + 5$$

$$\frac{200}{Q} = \frac{1}{10}Q$$

$$Q^2 = 2000$$

$$Q = \pm 44{\cdot}72,$$

the same answer as before.

(ii) *Profit Maximization*

Profits are defined as the excess of total revenue over total costs:

$$P = R - C$$

The rules for finding a maximum point tell us that P is maximized when the derivative of the profit function is equal to zero (and the second derivative is negative). If we denote the derivatives of the revenue and cost functions by dR and dC we have:

P is at a maximum when

$$dR - dC = 0.$$

This equation may also be written

$$dR = dC$$

The derivative of the total revenue function must be equal to the derivative of the total cost function for profits to be maximized. But we know that this is equivalent to saying that the best profit point occurs when marginal revenue is equal to marginal cost. This is a familiar economic proposition, for if marginal revenue is greater than marginal cost then the firm should expand production to collect the profit on the marginal item; if marginal revenue is less than marginal cost the extra item is being produced at a loss and the firm should cut back production.

Example

Find the best profit point of a firm whose total revenue and total cost functions are as follows:

$$R = 260Q - 3Q^2$$
$$C = 500 + 20Q$$

$$\begin{aligned}\text{Profit} &= R - C\\ &= 260Q - 3Q^2 - (500 + 20Q)\\ &= -3Q^2 + 240Q - 500\end{aligned}$$

Profit is maximized when

$$\begin{aligned}\frac{d}{dQ}(-3Q^2 + 240Q - 500) &= 0\\ -6Q + 240 &= 0\\ Q &= 40\end{aligned}$$

The optimum output for the firm will be 40 units of Q.

6. A Use of Integration

The problem of finding a suitable function to represent a cost curve may well involve the use of integration. For example, a firm will usually know its fixed costs. Let us suppose that the fixed or inescapable costs for a bookshop are Shs. 300 a month. The manager reckons that every book handled increases the costs of operation by 75 cents a year. Could we construct a suitable cost function for this shop in terms of books (B)?

The marginal cost per book for a month is $\frac{\cdot 75}{12} = \cdot 0625$ Shs.

The total cost function must be the integral of the marginal cost function:

$$\begin{aligned}MC &= \cdot 0625\\ C &= \textstyle\int \cdot 0625 \text{ with respect to B.}\\ &= \cdot 0625B + K \text{ (K is a constant).}\end{aligned}$$

The fixed costs of the business are Shs. 300 a month. Therefore K (which represents the value of C when $B = 0$) must be equal to 300. The complete function is

$$C = \cdot 0625B + 300$$

7. Elasticity

Elasticity of demand (or supply) is defined as the proportionate change in the quantity demanded or supplied divided by the proportionate change in price. If we let η denote elasticity

$$\eta = \frac{\frac{\Delta Q}{Q}}{\frac{\Delta P}{P}} = \frac{\Delta Q}{\Delta P} \cdot \frac{P}{Q}$$

By using the derivative $\frac{dQ}{dP}$ as the limiting value of $\frac{\Delta Q}{\Delta P}$ we have a very useful formula for calculating the elasticity at any point on a demand (or supply) curve.

$$\eta = \frac{dQ}{dP} \cdot \frac{P}{Q}$$

It should be noted that a demand curve will normally have negative elasticity: the amount demanded decreases as the price rises; a normal supply curve will have positive elasticity.

Example

Given the demand equation

$$Q = 150 - 15P;$$

compute elasticity when P = 4 and P = 8.

When P = 4,

$$Q = 150 - 15(4) = 90$$

$$\frac{dQ}{dP} = -15$$

$$\eta = \frac{dQ}{dP} \cdot \frac{P}{Q}$$

$$= (-15)\ \frac{4}{90}$$

$$= -\frac{2}{3}$$

When P = 8

$$Q = 150 - 15(8) = 30$$

$$\frac{dQ}{dP} = -15$$

$$\eta = \frac{dQ}{dP} \cdot \frac{P}{Q}$$

$$= (-15)\ \frac{8}{30}$$

$$= -4$$

We are normally interested in elasticity of demand because we wish to know how total expenditure will change when the price of a commodity rises or falls. The relationship between expenditure, price changes and elasticity can be shown as follows. If E denotes expenditure, P price and Q quantity, then E = P.Q. To find the effect upon expenditure when there is a change in price, we differentiate the function with respect to P, using the product rule:

$$\frac{dE}{dP} = Q + \frac{dQ}{dP} \cdot P$$

$$= Q(1 + \frac{dQ}{dP} \cdot \frac{P}{Q})$$

But since $\eta = -\frac{dQ}{dP} \cdot \frac{P}{Q}$, we have

$$\frac{dE}{dP} = Q(1 - \eta).$$

If $\eta > 1$, $\frac{dE}{dP}$ is negative, and expenditure will fall when price rises, rise when price falls. If $\eta = 1$, there is no change in expenditure when the price changes. If $\eta < 1$, $\frac{dE}{dP}$ is positive, and expenditure will rise when price rises, fall when price falls.

EXERCISE 10

The Production Function

1. Make a rough sketch of the following undefined functions and determine which would be suitable for representing a production function:

   (a) $y = 20 + 60x - x^2$

   (b) $y = x^2 - 50x$

   (c) $y = 16x + 3x^2 - \frac{x^3}{3}$

   (d) $y = \frac{x^3}{3} - 3x^2 + 16x$

   (e) $y = 40x - x^2$.

   What assumptions have you made about the characteristic shape of a production function? What do x and y stand for in the equations you have selected?

2. (a) A farmer knows that he can grow 50 pounds of cotton on his farm without using any fertilizer. The agricultural extension officer assures him that for each wheelbarrow-full of fertilizer he uses his crop will increase by 4 pounds. Find an equation to represent the production function of this farm.

   (b) A neighbouring farmer has been using fertilizers for some years and has come to the conclusion that whereas output can undoubtedly be raised from the minimum 50 pounds by the use of fertilizer, there is a limit to this process. He finds that after the 24th load the cotton crop actually begins to decline. Find a suitable equation (from among many possibilities) describing this farmer's cotton production function.

3. $0 = 10 + 12L - L^2$

   The above expression shows how total output O varies with the input of man-hours L.

   (a) Give an algebraic expression for
      (i) average product
      (ii) marginal product.

   (b) (i) How many man-hours of labour should be employed to maximize total output?
      (ii) What is the value of total output at this point?
      (iii) What is the value of the marginal product at this point?
      (iv) What is the value of the average product at this point?

4. Given the production function

$$O = 8 + 16F - F^2$$

   where O = total output, and F = units of input:

   (a) Use the expression $\triangle O \approx \frac{dO}{dF}(\triangle F)$ to calculate the approximate increase in total output as the amount of input is increased from 6 to 6·2

   (b) Check your result by substituting first 6, and then 6·2 in the original expression. How large is the error in (a)?

5. $O = 40F + 3F^2 - \frac{F^3}{3}.$

   The above expression shows how total output O varies with the input F.

   (a) Give an algebraic expression for the average product, and the marginal product.
   (b) Find the values of F and O for which total output is at a maximum.
   (c) Find the maximum value of the marginal product.
   (d) When do diminishing returns set in?
   (e) Draw a graph to illustrate your answer.

6. The employment of "a" man-days on "b" acres of land gives

$$O = 2(12ab - 5a^2 - 4b^2) \text{ bushels of wheat.}$$

   (a) Derive equations in terms of a and b for the marginal product of labour and the marginal product of land.

(b) If a man-day of labour costs the same as a day's rent for an acre of land, what will be the optimal proportion of labour to land to employ?

(c) Give the complete differential of this function.

(d) Use (c) to calculate the effect on total output if the number of man-days rises from 4 to 4·1 and the land acreage declines from 6 to 5·2.

7. A businessman, interested in maximizing the output of his firm wonders whether to employ male or female workers. Women he can employ for Shs. 25 a week; men for Shs. 50 a week. He estimates that he can spend Shs. 2000 a week on wages.

(a) Express in the form of an equation with variables M (number of male workers) and F (number of female workers), the possibilities open to the employer, given this expenditure constraint.

(b) Draw the curve of this equation, showing F on the vertical axis and M on the horizontal axis. What is the economic meaning of the slope of this curve?

Past experience has told this *entrepreneur* that he can obtain the same output using different combinations of men and women. The combinations of workers which produce the same output (O) may be represented

$$M \times F = O$$

Three levels of output which can be achieved by combining workers in this way are 200 units of output, 450 units of output and 800 units of output.

(c) Sketch these three isoquants (again measuring F on the vertical axis, and M on the horizontal axis). Explain the economic meaning of the slope of these functions.

Naturally, the businessman would like to produce the highest output possible—the one which is just touched by the expenditure constraint. When a straight line "just touches" a curve it is said to be tangential to it; at the point of tangency the line and the curve will have the same slope.

(d) What is the economic significance of the point of tangency?

(e) Find the points on the three isoquants which have the same slope as the employment constraint.

(Suggestion: express all equations in the form $F = f(M)$ and look for $\frac{dF}{dM}$).

(f) Which of these points lies on the constraint line? How many men and how many women should be employed to maximize output and what will output be?

Cost and Revenue Functions

8. The total cost function of a cotton ginnery is given by

$$C = \frac{1}{5}x^2 + 6x + 100$$

where x represents 100 lbs. of cotton.

(a) Draw the average and marginal cost functions on the same diagram showing values of x from 0 to 80.

(b) From the graph, find the value of x for which average cost is at a minimum. When is average cost equal to marginal cost?

9. (a) If the total revenue function for a firm is

$$R = f_1(Q)$$

and the total cost function is

$$C = f_2(Q)$$

explain

(i) the two algebraic conditions for maximum net revenue (= profits).

(ii) the economic meaning of these two conditions.

(b) Find the optimum output of a firm whose total revenue and total cost functions are given by

$$R = 30Q - Q^2$$
$$C = 20 + 4Q$$

Sketch these two functions.

10. (a) A cement works produces x tons of cement per week at a total cost of

$$£\ (\frac{1}{10}x^3 - 3x^2 + 50x + 300).$$

If the market price is fixed at £33⅓ per ton, what output per week will maximize net revenue?

(b) Explain carefully how an increase in fixed costs will affect a firm's optimum output. For example, if the fixed costs in the preceding question rise from 300 to 305 (and the market price is unaltered), will the cement works produce more, less or the same amount of cement as before?

11. The following functions show how C, total cost (in shillings) and R, total revenue (also measured in shillings) vary with output x (measured in 100's of units),

$$C = \frac{x^3}{3} - 3x^2 + 9x + 16$$

$$R = 21x - x^2$$

(a) Does this producer sell in a monopolistic or a perfectly competitive market?

(b) At what output does he maximize his total revenue? What is his total revenue at this point?

(c) When is his marginal cost at a minimum?

(d) What output will maximize this firm's total profit?

(e) Sketch the total cost and total revenue functions on the same diagram using this information.

Elasticity

12. The following diagram shows the demand curve for two different income groups and the supply curve for maize-meal in Uganda.

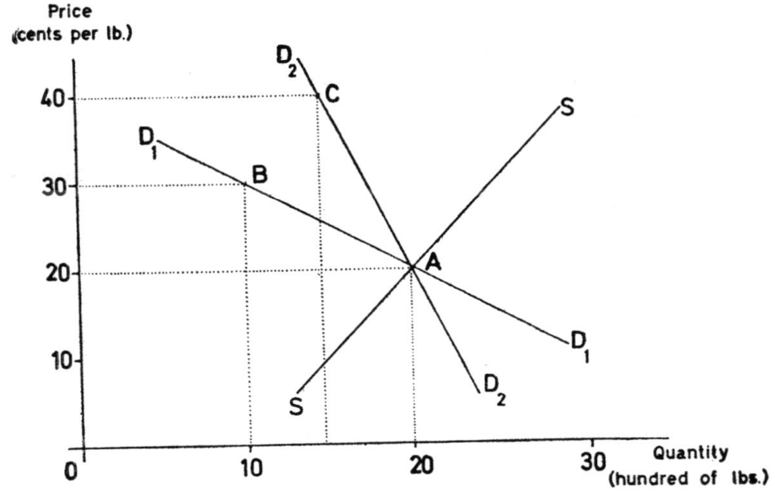

(a) What is the value of the price elasticity of demand for D_1 and D_2 and the price elasticity of supply at point A?
(b) What is the elasticity of D_1 at point B? of D_2 at point C?

13. (a) Explain the economic significance of the following demand elasticities :
    (i) $-\ 4$
    (ii) $-\ \infty$
    (iii) $-\ \frac{1}{8}$
    (iv) $-\ 1$
    (v) $0.$

(b) Given the demand equation

$$Q = \frac{90}{P}$$

compute elasticity when

(i) $P = 90$
(ii) $P = 30$
(iii) $P = 10$

(c) Given the supply equation

$$Q = 20 - \cdot 05P + P^{\frac{1}{2}}$$

compute elasticity when

(i) $P = 4$
(ii) $P = 16$
(iii) $P = 100.$

14. (a) Given

$$e = \frac{\frac{dQ^s}{Q^s}}{\frac{dP}{P}} \quad \text{and} \quad \eta = \frac{\frac{dQ^d}{Q^d}}{\frac{dP}{P}}$$

where e measures elasticity of supply and η measures the elasticity of demand for a commodity, find the proportionate change in price that will follow from a rightwards shift in the supply curve caused by a 15% increase in output.

Suggestion : $$\frac{dQ^s}{Q^s} = e\frac{dP}{P} + \cdot 15$$

(b) If $e = 2$, and $\eta = -2\cdot 5$, what will the proportionate change in price be?

PART THREE
ARITHMETIC OF GROWTH

XI. LOGARITHMS

XII. RATES OF GROWTH
LOGARITHMIC AND
EXPONENTIAL FUNCTIONS

XIII. ARITHMETIC AND GEOMETRIC
SERIES

XIV. ECONOMIC APPLICATIONS
GROWTH AND PLANNING

XI. LOGARITHMS

1. Introduction

Most students will be familiar with the use of common logarithms to simplify cumbersome multiplication and division problems. Paragraph 2 below reviews this process for the benefit of those who have forgotten. The rest of this section is concerned with the theoretical basis for the use of logarithms. Logarithms are exponents and it may be useful here to summarize the rules governing the use of exponents:

$$a^m \cdot a^n = a^{m+n}$$

$$a^m \div a^n = a^{m-n}$$

$$(a^m)^n = a^{mn}$$

Also, by definition $a^0 = 1$

2. Common Logarithms

(i) *The Use of Logarithm Tables*

Common logarithms (see the first table in the appendix) have been constructed to the base 10. A logarithm is simply the power, or exponent, to which 10 has to be raised to equal any given number. Thus

$$\begin{aligned}
1000 &= 10^3 \text{ therefore } & \log_{10}1000 &= 3{\cdot}0000\\
100 &= 10^2 & \log_{10}100 &= 2{\cdot}0000\\
10 &= 10^1 & \log_{10}10 &= 1{\cdot}0000\\
1 &= 10^0 & \log_{10}1 &= 0{\cdot}0000\\
0.1 &= 10^{-1} & \log_{10}0.1 &= -1{\cdot}0000\\
& & \text{usually written } &\ \ \bar{1}{\cdot}0000\\
0.01 &= 10^{-2} & \log_{10}0.01 &= \bar{2}{\cdot}0000
\end{aligned}$$

(The table can of course be continued in both directions.)

A logarithm is composed of two parts—the number before the decimal point which is known as the characteristic, and the numbers following the decimal point, called the mantissa. The former may be determined by inspection. The above table shows that any number between 100 and 1000 will have a logarithm between 2 and 3; the characteristic will be 2. Similarly any number between 10 and 100 will have a logarithm between 1 and 2; the characteristic will be 1. Numbers between 1 and 10 will have a logarithm between 0 and 1; the characteristic will be 0. All numbers between 0 and 1 will have negative logarithms; for example the logarithm of ·37 which lies between 0·1 and 1, will have a characteristic of $\bar{1}$. The characteristic of ·037 would be $\bar{2}$ as the number lies between 0·01 and 0·1.

The part of the logarithm following the decimal point is found in the logarithm tables. The logarithm of the number 4768 may be determined as follows:

As the number lies between 1000 and 10,000 (i.e. between 10^3 and 10^4) the characteristic is 3. To determine the remainder of the logarithm look down the column at the extreme left of the table for the first two figures 47; the logarithm of 4768 will be along the horizontal row to the right of 47. The third figure of the wanted number is 6; look along the row to the number in the main body of the table under the column marked 6. The logarithm given here is .6776. Finally, for the fourth figure 8, look along the same row to the columns of figures at the extreme right of the table. Under 8, we find 7. This number is added to .6776 to give .6783, and we have the final result $\log_{10} 4768 = 3.6783$.

It should be noted that four figure logarithm tables can only be used for expressions with four significant figures. Thus the number 435,772 must be rounded to 435,800 before using the tables. Conversely, the solution of a problem which has been solved by logarithms will only contain four figures and will only be accurate to three figures as the fourth figure may be rounded, as in the above example.

When the logarithm of a number is known, the number itself may be found from the *anti-logarithm* tables. When using anti-logarithm tables it is important to remember that only those numbers *following* the decimal point should be looked up in the table. The number before the decimal point, the characteristic, shows the position of the decimal point in the final answer.

Example $\log_{10}N = 2{\cdot}3681$

$N = 233{\cdot}4$

(ii) *Using Logarithms for Calculation*

The reason why logarithms can be used to simplify calculations follows from the rules given for multiplication and division of numbers with the same base. Because

$$43 = 10^{1.6335}$$
$$\text{and } 725 = 10^{2.8603}$$
$$43 \cdot 725 = 10^{(1.6335 + 2.8603)}$$
$$= 10^{4.4938}$$

Using anti-logarithm tables,

$$10^{4.4938} = 31180.$$

This problem may be written as follows:

43 · 725	Nos.	Logs.
	43	1.6335
	725	2.8603
	31180	4.4938

Answer 31180 (approximately).

Similarly, in a division problem the exponents, or logarithms, are subtracted.

(iii) *Logarithms to Other Bases*

It is, of course, possible to construct logarithms to any base. For example, if 2 is the base,

$$1 = 2^0 \text{ therefore } \log_2 1 = 0$$
$$2 = 2^1 \qquad \log_2 2 = 1$$
$$4 = 2^2 \qquad \log_2 4 = 2$$
$$8 = 2^3 \qquad \log_2 8 = 3$$
$$16 = 2^4 \qquad \log_2 16 = 4$$
$$32 = 2^5 \qquad \log_2 32 = 5$$
$$64 = 2^6 \qquad \log_2 64 = 6$$
$$128 = 2^7 \qquad \log_2 128 = 7$$

The above table can be used to solve the multiplication problem 8 · 16 by using logarithms to the base 2. Since

$$2^3 \cdot 2^4 = 2^7 = 128$$

the problem can be shown thus:

Nos.	Logs.
8	3
16	4
128	7

It should now be clear that logarithms are simply exponents, and technically the base of the logarithm should always be shown.

$$\log_{10}43 = 1{\cdot}6335$$
$$\log_2 8 = 3$$

In practice, when using common logarithms the base 10 is often omitted.

3. General Rules for the Use of Logarithms

(i) *The General Expression of a Logarithm*

From the preceding sections we derive the general formula

If $N = a^x$

$\log_a N = x.$

Two special cases are

$$\log_a 1 = 0 \ (a^0 = 1)$$

$$\log_a a = 1 \ (N = a).$$

(ii) *Laws Governing the Use of Logarithms*

I. The logarithm of a product:
$$\log_a MN = \log_a M + \log_a N.$$

II. The logarithm of a quotient:
$$\log_a \frac{M}{N} = \log_a M - \log_a N.$$

III. The logarithm of a power:
$$\log_a N^p = p \log_a N.$$

IV. The change of base in a logarithm:
$$\log_a N = \log_b N \log_a b.$$

(iii) *Proofs of the Laws of Logarithms*

The proofs of the above laws are included because they provide useful practice in understanding the basic relationships between the numbers and their logarithms. The first three proofs follow directly from the three laws given for the use of exponents, the fourth proof follows from the third.

I. $\text{Log}_a MN = \log_a M + \log_a N.$

Proof

$$\text{Let } \log_a M = x \;;\; M = a^x$$
$$\log_a N = y \;;\; N = a^y$$
$$\text{Then } MN = a^x a^y = a^{x+y}$$
$$\log_a MN = x + y$$
$$= \log_a M + \log_a N.$$

II. $\text{Log}_a \dfrac{M}{N} = \log_a M - \log_a N$

Proof

$$\text{Let } \log_a M = x \;;\; M = a^x$$
$$\log_a N = y \;;\; N = a^y$$
$$\text{Then } \frac{M}{N} = \frac{a^x}{a^y} = a^{x-y}$$
$$\log \frac{M}{N} = x - y$$
$$= \log_a M - \log_a N.$$

III. $\text{Log}_a N^p = p \log_a N.$

Proof

$$\text{Let } \log_a N = x \;;\; N = a^x$$
$$\text{Then } N^p = (a^x)^p = a^{xp}$$
$$\log_a N^p = xp$$
$$= p \log_a N.$$

IV. $\text{Log}_a N = \log_b N \log_a b.$

Proof

$$\text{Let } \log_b N = x \;;\; N = b^x$$
$$\text{Then } \log_a N = \log_a b^x$$
$$= x \log_a b \text{ (by Law III)}$$
$$= \log_b N \log_a b.$$

4. Some Examples Using Logarithms

(i) *A Numerical Calculation*

$$\frac{372^3 \times \cdot 56}{457}$$

No.	Log.	Log.
372^3	2·5705 (x3)	7·7115
·56		$\bar{1}$·7482
		7·4597
457		2·6599
63060		4·7998

Answer 63060 (approximately).

(ii) *An Equation in which the Unknown is an Exponent*

$$\frac{3^x}{10} - 42 = 0$$

$$3^x = 420$$

$$x \log 3 = \log 420$$

$$x = \frac{\log 420}{\log 3}$$

$$= \frac{2 \cdot 6232}{\cdot 4771}$$

$$= 5 \cdot 498 \text{ (approximately).}$$

(iii) *Expressing Logarithms to a Different Base*

(a) Show that $\log_a b = \dfrac{1}{\log_b a}$.

Law IV (rearranged) gives:

$$\log_a b \log_b N = \log_a N$$

Let $N = a$

Then $\log_a b \log_b a = 1$

$$\log_a b = \frac{1}{\log_b a}.$$

This useful result shows that Law IV may be expressed in two different ways:

$$\log_a N = \log_b N \log_a b = \frac{\log_b N}{\log_b a}.$$

(b) Evaluate $\log_2 304$.
Using the result of (a)

$$\log_2 304 = \frac{\log_{10} 304}{\log_{10} 2}$$

$$= \frac{2 \cdot 4829}{\cdot 3010}$$

$$= 8 \cdot 249 \text{ (approximately).}$$

EXERCISE 11

A. Evaluate to three significant figures

1. $\dfrac{625 \times 3560}{58 \cdot 7}$

2. $67 \cdot 388 \div 1294$

3. $\cdot 0754 \times \cdot 68$

4. $(\cdot 7943)^4$

5. $\sqrt[3]{25 \cdot 63}$

B. 1. Without using tables show that

(a) $\frac{1}{2} \log \dfrac{25}{9} - \log \dfrac{15}{4} + \frac{2}{3} \log \dfrac{27}{8} = 0.$

(b) $\dfrac{\log \sqrt{27} + \log \sqrt{8} - \log \sqrt{125}}{\log 6 - \log 5} = \dfrac{3}{2}$

2. Calculate $\log_{10} 5$ and $\log_{10} 125$ given that $\log_{10} 2 = \cdot 3010$

3. If $\log_x 10 \cdot 24 = 2$, find x.

4. Evaluate

(a) $\log_{25} 3 \cdot 142$

(b) $\left(\dfrac{1}{12}\right)^{1 \cdot 405}$

5. Solve for x:

   (a) $\log (x^2 + 2x) = 0{\cdot}9031$

   (b) $(2{\cdot}4)^x = 0{\cdot}59$

6. Solve for x:

   (a) $2^{x^2} = 16^{x-1}$.

   (b) $3^{2x} = 5^{x+1}$.

XII. RATES OF GROWTH. LOGARITHMIC AND EXPONENTIAL FUNCTIONS

1. Simple and Compound Interest

Interest is the payment made for the use of money and may be paid at simple or compound rates.

(i) *Simple Interest*

If money is invested at simple interest, interest payments are made each year on the *original* investment. Therefore, if P shillings are invested at a rate of r for t years, A, the value of the investment at the end of the period, will be given by

$$A = P + Prt.$$

Example

Calculate the value of an investment of Shs. 500 after 3 years if simple interest is paid at an annual rate of 5%.

$$\begin{aligned} A &= P + Prt \\ &= 500 + (500 \cdot \frac{5}{100} \cdot 3) \\ &= 575 \text{ shillings.} \end{aligned}$$

(ii) *Compound Interest*

If money is invested at compound interest, interest payments are made each year not on the original sum, but on the *current value* of the original investment, i.e. the original investment plus the accumulated interest. If P shillings are invested at compound interest at a rate r, the amount A at the end of any given period may be calculated as follows:

after one year

$$\begin{aligned} A &= P + Pr \\ &= P(1 + r) \end{aligned}$$

after two years

$$\begin{aligned} A &= P(1 + r) + r\,P(1 + r) \\ &= P(1 + r)(1 + r) \\ &= P(1 + r)^2 \end{aligned}$$

after three years

$$\begin{aligned} A &= P(1 + r)^2 + r\,P(1 + r)^2 \\ &= P(1 + r)^2\,(1 + r) \\ &= P(1 + r)^3 \end{aligned}$$

after t years

$$A = P(1 + r)^t.$$

Example
Calculate the value of an investment at the end of 6 years if Shs. 1000 is invested at 4% p.a. compound interest

$$\begin{aligned} A &= P(1 + r)^t \\ &= 1000\;(1 + \cdot 04)^6 \\ &= 1000\;(1 \cdot 04)^6 \end{aligned}$$

Taking logs

$$\begin{aligned} \log A &= \log 1000 + 6 \log 1 \cdot 04 \\ &= 3 + 6\;(\cdot 0170) \\ &= 3 \cdot 1020 \\ A &= 1265 \text{ shillings.} \end{aligned}$$

If interest is paid more than once a year, the value of the investment will rise even more rapidly. If, for example, interest is paid every 6 months, there will be two interest payments a year, but at one half the annual rate. The compound interest formula becomes

$$A = P\left(1 + \frac{r}{2}\right)^{2t}$$

and in general if interest is paid n times a year

$$A = P\left(1 + \frac{r}{n}\right)^{nt}.$$

Example
If Shs. 1000 is invested at 4% p.a. compound interest and interest payments are made four times a year, what will be the value of the investment at the end of 6 years?

$$\begin{aligned} A &= P\left(1 + \frac{r}{n}\right)^{nt} \\ &= 1000\left(1 + \frac{\cdot 04}{4}\right)^{4 \cdot 6} \\ &= 1000\;(1 \cdot 01)^{24} \end{aligned}$$

Taking logs

$$\log A = \log 1000 + 24 \log 1{\cdot}01$$
$$= 3 + 24({\cdot}0043)$$
$$= 3{\cdot}1032$$
$$A = 1269 \text{ shillings.}$$

This result should be compared with that obtained in the previous example.

(iii) *Discounted Values and Capital Values*

In the previous examples the present value of an investment was known and the problem was to calculate its future value. It is often useful to be able to find the present value of an asset which will be worth a certain amount at some future date. This process is known as discounting. In terms of the compound interest formula A, the future value is known, but not P. Rearranging the formula

$$P = \frac{A}{(1 + r)^t}$$

Example

A government bond will be worth Shs. 1000 in ten years time. What is it worth today if 5% compound interest is assumed?

$$P = \frac{A}{(1 + r)t}$$
$$= \frac{1000}{(1{\cdot}05)^{10}}$$

Taking logs

$$\log P = \log 1000 - 10 \log 1{\cdot}05$$
$$= 3 - 10({\cdot}0212)$$
$$= 2{\cdot}7880$$
$$P = 613{\cdot}8 \text{ shillings.}$$

It is also possible to calculate the present value of an asset if the annual income from that asset in future years is known. This is known as the capital value of the asset.

If the current rate of interest is r, and the income from an asset in successive years is given by

$$a_0, a_1, a_2, \ldots\ldots\ a_t$$

then the capital value of the asset is

$$a_0 + \frac{a_1}{1 + r} + \frac{\imath_2}{(1 + r)^2} \cdots\ + \frac{a_t}{(1 + r)^t}$$

Example

The value of a cotton crop in the present year and in the four following years from an acre of land is expected to be Shs. 160, Shs. 150, Shs. 150, Shs. 130, and Shs. 100. In the next year the owner of the land hopes to sell it for Shs. 800. What is the present or capital value of this acre of land if a 4% rate of interest is assumed? Capital value

$$= 160 + \frac{150}{1{\cdot}04} + \frac{150}{(1{\cdot}04)^2} + \frac{130}{(1{\cdot}04)^3} + \frac{100}{(1{\cdot}04)^4} + \frac{800}{(1{\cdot}04)^5}$$

$$= 160 + 144{\cdot}2 + 138{\cdot}7 + 115{\cdot}5 + 85{\cdot}51 + 657{\cdot}9$$

$$= 1301.81 \text{ shillings.}$$

2. "e"

There is a hypothetical case of compound interest which has been of great interest to mathematicians. It was noted above that if interest payments are made more than once a year, the value of an investment tends to increase more rapidly. It is interesting to consider how rapidly the value of an investment will grow if interest payments are made more and more frequently. To simplify the arithmetic, assume that one shilling is invested for one year at 100%. In this case $P = 1, r = 1, t = 1$. Assuming that interest is paid n times a year the compound interest formula

$$A = (1 + \frac{r}{n})^{nt}$$

becomes $A = (1 + \frac{1}{n})^{n}$

When n = 1 A = 2
n = 2 A = 2·25
n = 3 A = 2·37
n = 4 A = 2·44

. . . .

n = 10 A = 2·594
n = 100 A = 2·704
n = 1,000 A = 2·717
n = 10,000 A = 2·718

Clearly, as n increases, A appears to be tending to a limit, and in fact it is the case that as n increases the value of $(1 + \frac{1}{n})^n$ tends to a limit of 2.71828. This limiting value is denoted by the letter "e". Formally

$$e = \text{limit } (1 + \frac{1}{n})^n \text{ as } n \to \infty.$$

In the above example e is the value to which an investment of one shilling will tend if 100% interest is paid continuously for one year. The application of "e" in mathematics is much wider than this hypothetical compound interest problem, and in particular e is frequently used as a logarithmic base. Logarithms to the base e are known as *natural logarithms* and may be found in the appendix under the heading "Napierian logarithms". These tables may be used in exactly the same way as common logarithms to the base 10.

3. Graphs of Logarithmic and Exponential Functions

Below is the graph of $y = \log_{10}x$, showing values of x from 1 to 8.

x	1	2	3	4	5	6	7	8	9	10
y	·0000	·3010	·4771	·6021	·6990	·7782	·8451	·9031	·9542	1·000

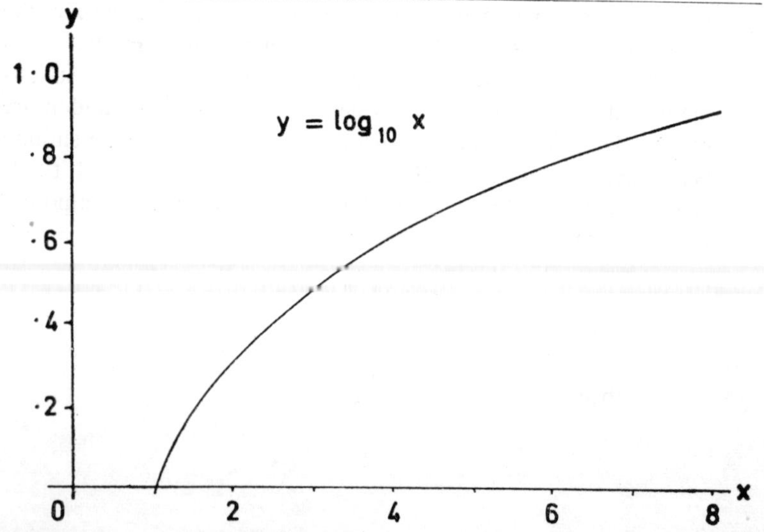

This graph has the characteristic shape of a *logarithmic function,* showing that constant absolute increases in x lead to a declining rate of growth in y. The reason for this is that logarithmic scales reflect proportional changes. Thus a change in x from 1 to 2 represents an increase of 100%, but the increase from 2 to 3 represents an increase of only 50%. Similarly the increase from 3 to 4 is only $33\frac{1}{3}$% and so the graph of $y = \log_{10}x$ shows a declining growth rate.

The next graph shows the characteristic shape of an *exponential function*—a function where the independent variable is an exponent. The graph shows $y = 10^x$ for values of x from 0 to 1.

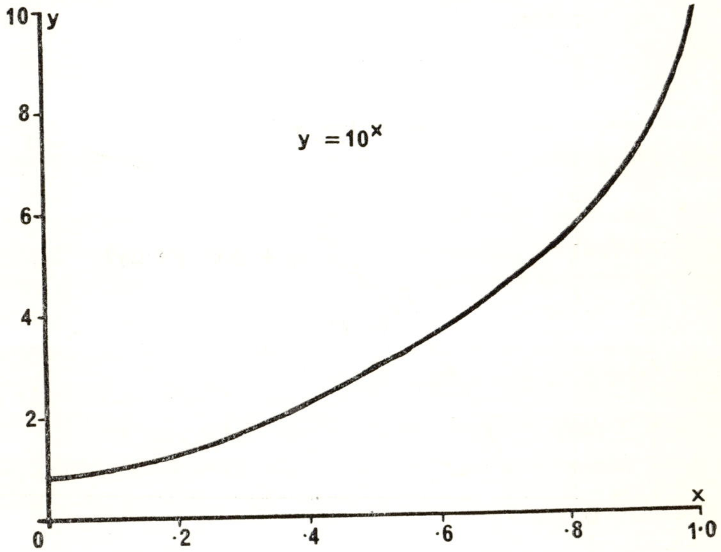

This graph is the converse of the one above. From the preceding Chapter we know that $y = \log_{10}x$ is equivalent to $x = 10^y$, so to draw the graph of $y = 10^x$ it is only necessary to reverse the axes.

The graph of an exponential function shows that constant increases in the independent variable (x) lead to rapidly rising increases in y, and it should be noted that growth rates such as

$$A = P(1 + r)^t$$

are exponential functions of t. As t increases, the successive increments in A increase rapidly in size.

4. Differentiation of Logarithmic and Exponential Functions

(i) *The Derivative of* $y = \log_e x$.

For purposes of calculus, it is common to use a logarithmic

function to the base e. The standard form of the derivative of the function

$$y = \log_e x$$

is given by

$$\frac{dy}{dx} = \frac{1}{x}$$

The proof of this standard form is included as a useful exercise in both differentiation and the theory of logarithms.

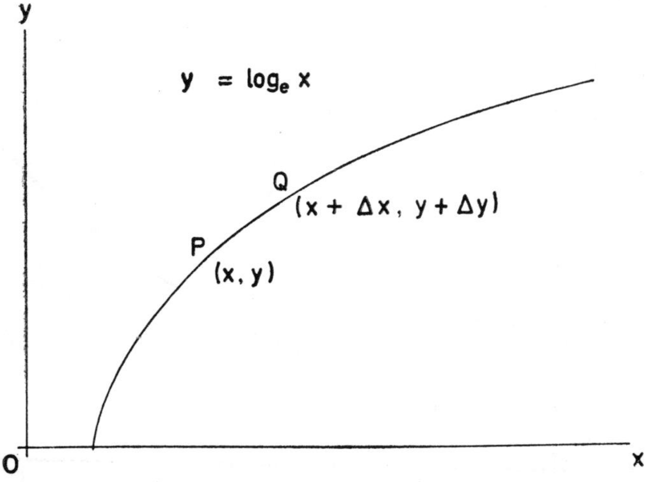

Proof

$$y = \log_e x.$$

Let P(x,y) and Q(x + $\triangle$x, y + $\triangle$y) be points on the above curve.

At Q, $y + \triangle y = \log_e (x + \triangle x)$

At P $\quad y = \log_e x$

$$\triangle y = \log_e (x + \triangle x) - \log_e x$$

$$= \log_e \frac{(x + \triangle x)}{x}$$

$$= \log_e (1 + \frac{\triangle x}{x})$$

$$\frac{\triangle y}{\triangle x} = \frac{1}{\triangle x} \log_e (1 + \frac{\triangle x}{x})$$

$$= \log_e \left(1 + \frac{\Delta x}{x}\right)^{\frac{1}{\Delta x}}$$

$$= \frac{1}{x} \log_e \left(1 + \frac{\Delta x}{x}\right)^{\frac{x}{\Delta x}}$$

Let $\dfrac{x}{\Delta x} = n$

Then $\dfrac{\Delta y}{\Delta x} = \dfrac{1}{x} \log_e \left(1 + \dfrac{1}{n}\right)^n$

As $Q \rightarrow P$

$$\Delta x \rightarrow 0$$

$$n \rightarrow \infty$$

$$\frac{\Delta y}{\Delta x} \rightarrow \frac{dy}{dx} = \frac{1}{x} \log_e e = \frac{1}{x}$$

(ii) *Derivative of an Exponential Function*
The standard form for the derivative of

$$y = e^x$$

is $\dfrac{dy}{dx} = e^x$

Proof

$$y = e^x$$

Then $x = \log_e y$

$$\frac{dx}{dy} = \frac{1}{y}$$

and $\dfrac{dy}{dx} = y = e^x$

(iii) *Some Examples of Logarithmic and Exponential Derivations*
The two standard forms proved above are often used in conjunction with the function of a function rule (see p. 61).

This gives the following results:

If $y = \log_e u$ (where $u = f(x)$)

$$\frac{dy}{dx} = \frac{dy}{du} \cdot \frac{du}{dx} = \frac{1}{u} \cdot \frac{du}{dx}$$

If $y = e^u$ (where $u = f(x)$)

$$\frac{dy}{dx} = \frac{dy}{du} \cdot \frac{du}{dx} = e^u \cdot \frac{du}{dx}$$

Examples

$$y = \log_e (x + 1), \qquad \frac{dy}{dx} = \frac{1}{x + 1}.$$

$$y = \log_e (2x - 3), \qquad \frac{dy}{dx} = \frac{2}{2x - 3}$$

$$y = \log_e x^2, \qquad \frac{dy}{dx} = \frac{2x}{x^2} = \frac{2}{x}$$

$$y = e^{2x}, \qquad \frac{dy}{dx} = 2e^{2x}$$

$$y = e^{1-3x}, \qquad \frac{dx}{dx} = -3e^{1-3x}$$

(iv) *The Integrals of* $\frac{1}{x}$ *and* e^x

The general form of the indefinite integral

$$\int (ax^n)dx = \frac{ax^{n+1}}{n + 1} + C$$

cannot be used for the integral

$$\int \frac{1}{x} \cdot dx$$

since an expression with a denominator of 0 would be obtained. It is therefore useful to have discovered

$$\int \frac{1}{x} \cdot dx = \log_e x + C.$$

The integral of e^x is simple,

$$\int e^x \cdot dx = e^x + C$$

$y = e^x$ being, as proved above, the only function which is equivalent to its own derivative.

EXERCISE 12

PART A

1. If Shs. 2000 is invested at 6% p.a. how much will it be worth in 8 years' time?

(a) At simple interest
(b) At compound interest.

2. If Shs. 1500 is invested at 6% p.a. and compound interest payments are made every month, how much will the investment be worth in 5 years' time?

3. Find the amount at the end of 10 years when Shs. 400 is invested at 4% compound interest
   (a) if interest is added annually
   (b) if interest is added twice a year.

4. Find what sum a man has to invest on his fortieth birthday so that he may be able to draw out a lump sum of Shs. 2000 on his sixtieth birthday, the investment being made at 5% p.a. compound interest.

5. A Shs. 5000 Government bond is due to mature in twelve years time. What is it worth today if the current rate of interest is 5%
   (a) if simple interest is assumed
   (b) if compound interest is assumed?

6. Assess the capital value of shares in a trading company which are expected to bring in dividends Shs. 40, Shs. 60, Shs. 90 and Shs. 100 in the present and the three following years, and which, it is hoped, will sell for Shs. 1000 in the fifth year
   (a) if the annual rate of interest is 6%
   (b) if the annual rate of interest is 3%.

PART B

7. Using Napierian logarithms to the base e, evaluate
   (a) 837×92
   (b) $213 \div 3{\cdot}7$

8. Draw the graph of $y = 2^x$ showing values of x from -2 to $+3$. Hence, draw a second graph of $y = \log_2 x$.

9. Describe the shape of the function that would result if the logarithms of the following series of numbers were plotted on a graph
   (a) 3, 6, 9, 12, 15 . . . .
   (b) 2, 4, 8, 16, 32 . . . .
   Check your answer by looking up the logarithms of the above numbers.

10. Use logarithm tables to construct a graph of $y = \log_{10}x$. Add the graph of $y = x - 2$ and so solve $\log_{10}x + 2 = x$ approximately.

PART C

11. Differentiate:
   (a) $y = e^{x+1}$
   (b) $y = e^{x^2 - 3x}$
   (c) $y = \frac{1}{e^x}$
   (d) $y = (3x^2 - 3e^x)^2$

12. Differentiate:
   (a) $y = \log_e (2x + 1)$
   (b) $y = \log_e (7 - x^2)$
   (c) $y = \log_e x + 3 \log_e x$
   (d) $y = 2 \log_e x - \log_e x^2$.

13. Integrate:
   (a) $\int (3e^{3x-1})dx$
   (b) $\int (2e^{2x} - 4x)\, dx$
   (c) $\int \left(\frac{7}{7x - 3}\right) dx$
   (d) $\int_1^{10} \left(\frac{1}{x}\right) dx$.

XIII.
ARITHMETIC AND GEOMETRIC SERIES

1. Series

A set of numbers which can be obtained from some definite law is called a series or a progression. Each of the numbers forming the set is called a term of the series.

Examples

(a) 1, 4, 7, 10 .....

(b) 2, 4, 8, 16 ......

(c) 1, 4, 9, 16 ......

In the first set, each number is obtained by adding 3 to the preceding one; in the second each term is twice the preceding one, and in the third each number is the square of successive integers (1^2, 2^2, 3^2, 4^2 ....). In each of the series the nth term may be deduced from the general law governing the series. The nth term of (a) is 3n—2; the nth term of (b) is 2^n, and the nth term of (c) is n^2.

Two simple types of series are the arithmetic and the geometric progression.

2. Arithmetic Progressions

An arithmetic progression is a series in which each term is obtained from the preceding one by the addition (or subtraction) of a constant quantity. This constant quantity is known as the common difference (d).

600, 650, 700, 750 . . . (d = 50)

8, 12, 16, 20, ... (d = 4)

14, 9, 4, −1, ... (d = −5)

The general expression may be given by

$$a,\ a + d,\ a + 2d,\ a + 3d \ldots\ldots a + (n - 1)\,d.$$

The sum of an arithmetic progression may be obtained as follows

Let S_n = the sum to n terms
a = the first term
d = the common difference
p = the last term

Then

$$S_n = a + (a + d) + (a + 2d) \dots + (p - d) + p$$

Reversing the order:

$$S_n = p + (p - d) + (p - 2d) \dots + (a + d) + a$$

By addition

$$2S_n = (a + p) + (a + p) + (a + p) \dots + (a + p)$$
$$= n\,(a + p)$$

$$S_n = \frac{n}{2}(a + p).$$

This general expression for the sum of an arithmetic progression may also be written

$$S_n = \frac{n}{2}(2a + (n - 1)\,d)$$

by substituting $a + (n - 1)\,d$ for p, the last term.

Example

Sum the series 7, 14, 21, 28 ..... to twenty terms.

$$a = 7$$
$$d = 7$$
$$n = 20.$$

$$Sn = \frac{n}{2}\Big[2a + (n - 1)d\Big]$$
$$= \frac{20}{2}\Big[14 + (19)(7)\Big]$$
$$= 1470.$$

3. Geometric Progressions

A geometric progression is a series in which each term is obtained from the preceding one by multiplying (or dividing) by a constant quantity. This constant quantity is known as the common ratio (r).

3, 6, 12, 24 . . . . . $(r = 2)$

1, −3, 9, −27 . . . . . $(r = -3)$

16, 4, 1, $\frac{1}{4}$ . . . . . $(r = \frac{1}{4})$.

The general expression may be given by

$$a,\ ar,\ ar^2,\ ar^3 \ \dots \ ar^{n-1}.$$

The sum of a geometric progression may be obtained as follows:

Let S_n = the sum to n terms
a = the first term
r = the common ratio.

Then

$$S_n = a + ar + ar^2 \quad . \quad . \quad ar^{n-2} + ar^{n-1}$$

Multiplying by r:

$$r\,S_n = ar + ar^2 \quad . \quad . \quad . \quad ar^{n-2} + ar^{n-1} + ar^n$$

by subtraction

$$S_n - rS_n = a - ar^n$$

$$S_n = \frac{a\,(1 - r^n)}{1 - r}$$

Example
Sum the geometric progression
1, –4, 16, –64 . . . . to six terms

$a = 1$
$r = -4$
$n = 6$

$$S_n = \frac{a\,(1 - r^n)}{1 - r}$$

$$= \frac{1 - (-4)^6}{1 - (-4)}$$

$$= \frac{1 - 4096}{5}$$

$$= -\ 819.$$

4. Divergent and Convergent Series

There are some geometric series in which each term is larger than the preceding one.

$$1 + 3 + 9 + 21 + 81 \ldots.$$

$$1 - 3 + 9 - 21 + 81 \ldots.$$

Clearly the greater the number of terms taken, the larger will be the sum of the series. Such a series is said to be *divergent.*

In other geometric series each term is smaller than the preceding one.

$$1 + \tfrac{1}{2} + \tfrac{1}{4} + \tfrac{1}{8} + \tfrac{1}{16} \; \ldots\ldots$$

$$1 - \tfrac{1}{2} + \tfrac{1}{4} - \tfrac{1}{8} + \tfrac{1}{16} \; \ldots\ldots$$

In this case, we can continue adding terms indefinitely, but soon each additional term will become so small that it will make very little difference to the sum of the series. Such a series is said to be *convergent*. A geometric progression will be convergent if the common ratio is a fraction; more formally, the series

$$a, ar, ar^2, \; \ldots\ldots$$

will be convergent if $-1 < r < 1$.

The *sum* of a convergent series tends to a limit. The sum of the above series
$1 + \frac{1}{2} + \frac{1}{4} + \frac{1}{8} \ldots\ldots$ is clearly tending to a limit of 2 for no matter how many terms are added the sum of the series will always differ from 2 by the value of the last term taken, for example,

$$1 + \tfrac{1}{2} + \tfrac{1}{4} + \tfrac{1}{8} = 1\tfrac{7}{8} = 2 - \tfrac{1}{8}.$$

In the case of a convergent series the formula for the sum of a geometric progression

$$S_n = a\frac{(1 - r^n)}{1 - r}$$

becomes, as $n \to \infty$

$$S_n \to \frac{a}{1 - r}$$

The term r^n becomes smaller as larger values of n are taken, and approaches zero. The expression $\frac{a}{1 - r}$ represents the *limit* of the sum of the series.

In the series $1 + \frac{1}{2} + \frac{1}{4} + \frac{1}{8} \ldots\ldots$

$$a = 1$$
$$r = \tfrac{1}{2}$$

and the limit of $S_n = \dfrac{1}{1 - \frac{1}{2}} = 2.$

Example

Show that the sum of the series.

$$1, \; -\tfrac{1}{3}, \; \frac{1}{9}, \; -\frac{1}{27} \; .....$$

tends to a limit of $\frac{3}{4}$.

The common ratio $-\frac{1}{3}$ lies between -1 and 1, therefore the series is convergent.

$$\text{The limit of } S_n = \frac{a}{1 - r}$$

$$= \frac{1}{1 - (-\frac{1}{3})}$$

$$= \frac{1}{1\frac{1}{3}}$$

$$= \tfrac{3}{4}.$$

EXERCISE 13

1. Write down the first three and the eighth term of the series whose nth terms are:-

   (a) $4n - 5$

   (b) 3^{n-1}

   (c) $(-1)^n$.

2. Find the sum of ten terms of an arithmetic progression of which the first term is 60 and the last is -104.

3. Sum the series
   $(-4) + (-1) + 2 + 5 \; .....$ to 20 terms.

4. Insert six terms in an arithmetic progression between 2 and 30.

5. Find three numbers in arithmetic progression such that their sum is 27 and their product is 504.

   (Suggestion : Let $a - d$, a and $a + d$ be the numbers).

6. Sum the geometric progression
   $2 + (-4) + 8 + (-16) \; .....$ to eight terms.

7. Find three numbers in geometric progression such that their sum is 39 and their product is 729.

   (Suggestion : Let the numbers be $\frac{a}{r}$, a, ar).

8. Show that the series

$$1 + \tfrac{3}{4} + 9/16 + 27/64 + \ldots\ldots$$

is convergent and find the limit of its sum.

9. Find the limit of the sum of the series

$$1000 + 800 + 640 \ldots\ldots$$

10. When does the series

$$a + \frac{a}{1 + a} + \frac{a}{(1 + a)^2} + \ldots\ldots$$

converge, and what is then the limit of its sum?

XIV. ECONOMIC APPLICATIONS: GROWTH AND PLANNING

1. Growth Rates and Development Plans

Economic growth rates are always calculated on a compound interest basis. A moment's reflection shows that this is reasonable. The growth of one year forms part of the basis for the following year's growth and the typical pattern of growth of an expanding economy will be an exponential function of time.

$$Y_t = Y_o (1 + r)^t$$

where Y_o = income in the base year

t = the number of years considered

Y_t = income after t years

r = the annual rate of growth.

The graph of this function will have the following shape:

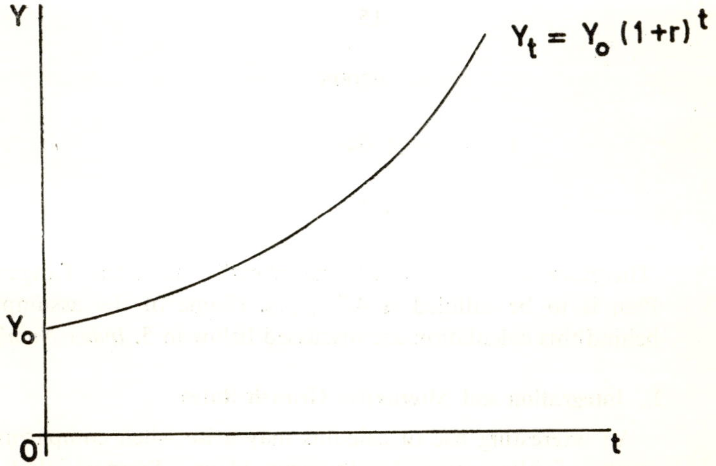

In May, 1964, a five year Development Plan was introduced in Tanganyika (now Tanzania). One of its objectives was to double real *per capita* income over the next 15 years. What is the average annual rate of growth required to achieve this objective?

At first glance, this problem seems to contain too many unknowns for us to use the growth formula

$$Y_t = Y_o (1 + r)^t.$$

We know t: the number of years is 15; the other known fact is that *per capita* income at the end of the period is to be twice the level of income at the beginning of the period:

$$Y_t = 2Y_O$$

Substituting in the equation gives

$$2Y_O = Y_O\,(1 + r)^{15}.$$

Dividing both sides by Y_O:

$$2 = (1 + r)^{15}.$$

Now we have one equation and one unknown. Solving by logarithms :

$$\log 2 = 15 \log(1 + r)$$

$$\log (1 + r) = \frac{\log 2}{15}$$

$$= \frac{\cdot 3010}{15}$$

$$= \cdot 02006$$

$$1 + r = 1 \cdot 047$$

$$r = 4 \cdot 7.$$

Therefore the required average growth rate if the Tanganyika Plan is to be fulfilled is 4.7% p.a. (Some of the assumptions behind this calculation are discussed below in 5. *Index Numbers.*)

2. Integration and Alternative Growth Rates

An interesting use of calculus may arise when comparing the merits of alternative development plans. Suppose that one set of experts recommends investing £45m. in the first year of a proposed 10 year Plan, and then plans to increase the level of investment (and therefore G.D.P.) by 6% p.a. The level of investment after a period of t years will be

$$I = 45(1.06)^t. \;\ldots\; \text{Plan 1}$$

A second set of advisors to the Government consider that a larger sum of money could be invested in the first year but they are less optimistic about the probable rate of growth. They postulate a function

$$I = 50(1.04)^t. \ldots\ldots \text{ Plan 2}$$

The Government has to weigh the benefits of these alternative plans. In the first years of the proposed plan, Plan 2 will require more funds for investment. By the end of the 10-year period, however, Plan 1 is generating £80.54m. a year in investment funds, while Plan 2 is providing £73.96m. in investment funds. Looking at the ten-year period as a whole, which Plan provides more funds for investment purposes?

The graph of the two plans look like this:

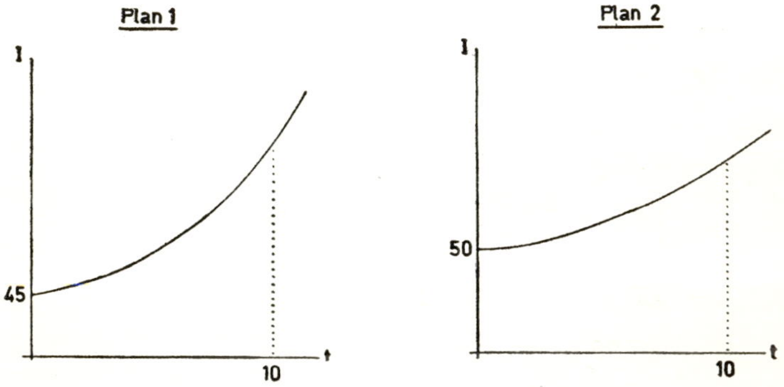

The total investment for all the ten years taken together must be represented by the areas under the curves. By integration we can sum the total area under the curve from the beginning of the Plan ($t = 0$) to the end ($t = 10$).

The particular functions we have been dealing with are

$$I_1 = 45(1.06)^t \quad . \quad . \quad . \quad . \quad \text{Plan (1)}$$
$$I_2 = 50(1.06)^t \quad . \quad . \quad . \quad . \quad \text{Plan (2).}$$

These functions can be expanded by the binomial theorem (see XV). Disregarding the higher powers of t, Plan (1) becomes

$$I_1 = 45 + 2{\cdot}619t + {\cdot}081t^2 + \ldots\ldots$$

and Plan (2) becomes

$$I_2 = 50 + 1{\cdot}96t + {\cdot}04t^2 + \ldots\ldots$$

Looking first at Plan (1)

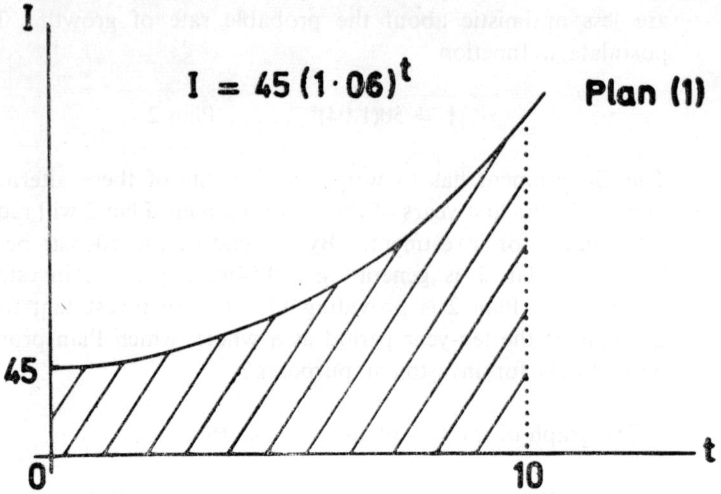

The area under the curve between t = 0, and t = 10 is given by

$$\int_0^{10} (45 + 2{\cdot}619t + {\cdot}081t^2 + \ldots .)\,dt$$

$$= \left[45t + \frac{2{\cdot}619t^2}{2} + \frac{{\cdot}081t^3}{3} + \ldots\right]_0^{10}$$

$$= 450 + 130{\cdot}95 + 27 + \ldots .$$

$$= 607.95 \text{ (approximately).}$$

Thus Plan (1) will lead to a total "self-generated" investment programme of £607.95m.

For Plan (2) the area under the curve between t = 0 and t = 10 is given by

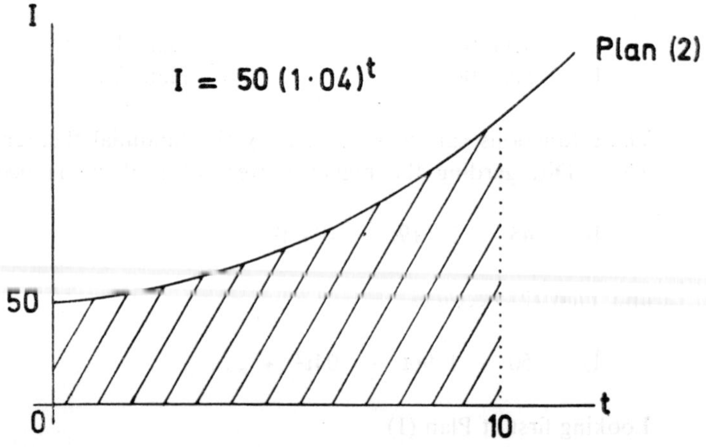

$$\int_0^{10} (50 + 1{\cdot}96t + .04t^2 + \ldots\ldots)\, dt$$

$$= \left[50t + \frac{1.96t^2}{2} + \frac{.04t^2}{3} + \ldots\ldots\right]$$

$$= 500 + 98 + 13{\cdot}3 + \ldots.$$

$$= 611{\cdot}3 \text{ (approximately).}$$

Plan (2) will lead to a total investment programme of £611·3m.

Over the ten year period Plan (2) provides rather more funds for investment. If a longer period were considered, however, the reverse would prove to be true.

3. The "Multiplier"

One aspect of economic growth —the "multiplier" analyis—is a well-known part of Keynesian theory. It is based on the mathematics of Chapter XIII.

Keynes maintained that in a given community people will tend to spend a certain proportion of any additional income they may receive, and save the rest. If by means of statistical research this "marginal propensity to consume" can be established, then an important economic consequence follows: an additional expenditure of £a in an economy will increase the income of the community by some multiple of a.

Let us suppose that a £10m. loan is made to the Kenya Government and that this sum is set aside for expenditure on road improvement. Keynes' argument was that as the roads were built, the £10m. passed into the hands of the contractors, the people who provide the stones and tar, the road-builders themselves. If the marginal propensity to consume is assumed to be $\frac{3}{4}$ in Kenya, then $\frac{3}{4}$ of this additional income will be spent by the contractors, workmen, etc., on such things as food, clothing, school-fees. This means additional income for the shopkeepers and education authorities. If they, in their turn, spend $\frac{3}{4}$ of this money, we have the beginning of a geometric progression

$$£\left[10 + \tfrac{3}{4}(10) + \tfrac{3}{4}(\tfrac{3}{4})(10) + \ldots.\right] \text{m.}$$

Provided the marginal propensity to consume remains constant at $\frac{3}{4}$, the original loan of £10m. will continue to grow. The common ratio of the geometric progression lies between —1 and

+ 1 so we know that the series is convergent and the sum of the series will tend to a limit.

$$S_n = \frac{a}{1 - r}$$

$$= \frac{10}{1 - \frac{3}{4}}$$

$$= £40\text{m.}$$

The original loan of £10m. has been multiplied by 4. Keynes, using the mathematics of the convergent geometric series defined the "multiplier" as $\frac{1}{1 - c}$ where c is the marginal propensity to consume.

There are many assumptions contained within this argument. An obvious one is that for any community there *is* a discoverable marginal propensity to consume out of additional income. Another even more fundamental assumption is that there are currently unemployed resources which can be used to produce new goods and services for people to spend their additional income on. Before accepting the multiplier analysis for East Africa the validity of these assumptions has to be examined.

4. The Use of Approximation in Calculating Growth Rates

Frequently in calculation of economic growth rates we come across problems that can only be solved by approximation. For example, suppose that 250 students were admitted to Makerere University College, Uganda, in 1964. Over the next fourteen years it is hoped to increase the annual intake so rapidly that the total intake over all fourteen years will be 25,500 students. What must be the annual rate of expansion if this target is to be achieved?

This looks like a straightforward problem. The original intake is 250 students. If we let the required rate of expansion be r the number of students entering Makerere over the successive years will be

$$250 + 250r + 250r^2 + 250r^3 + \ldots\ldots$$

We have a geometric progression of the form

$$a + ar + ar^2 + ar^3 + \ldots\ldots$$

The sum of a geometric progression (see p. 129) is given by

$$S_n = \frac{a\,(1 - r^n)}{1 - r}$$

Using the values of this problem (a = 250, n = 14, S_n = 25,500) we have

$$25{,}500 = 250\left(\frac{1 - r^{14}}{1 - r}\right)$$

$$102 = \frac{1 - r^{14}}{1 - r}.$$

We do not know any method of solving equations of the 14th degree! We cannot simply use logarithms because our unknown, r, appears in both numerator and denominator in the form of a compound expression. The only possible method of proceeding is to try a number of guesses for the value of r, and hope to achieve approximately the right solution. Let us take an increase of 20% p.a. for our first approximation. This means the rate of growth will be 120% p.a., r = 1·2.

$$\frac{1 - 1{\cdot}2^{14}}{1 - 1.2} = \frac{1 - 12{\cdot}84}{1 - 1.2} = \frac{-\ 11{\cdot}84}{-\ 0{\cdot}2} = 59{\cdot}2.$$

This answer is too low, i.e. below the 102 specified above; r must be greater than 120%. Now try r = 130% p.a.

$$\frac{1 - (1{\cdot}3)^{14}}{1 - 1{\cdot}3} = \frac{1 - 39{\cdot}31}{1 - 1{\cdot}3} = \frac{-\ 39{\cdot}31}{-\ {\cdot}3} = 127{\cdot}7$$

This answer is too high, i.e. above the 102 specified above; r must be less than 130%. However, the correct answer is clearly nearer 130% than 120%, so we try 127%.

$$\frac{1 - (1{\cdot}27)^{14}}{1 - 1{\cdot}27} = \frac{1 - 28{\cdot}4}{1 - 1{\cdot}27} = \frac{-\ 27{\cdot}4}{-\ 0{\cdot}27} = 101{\cdot}5$$

This answer is near enough for our purpose. Further approximation might change the answer to 127.1% or perhaps 127.15, but certainly not to 128%. The annual intake into Makerere must therefore grow at an annual rate of 27% (to the nearest per cent), if the total intake over the next fourteen years is to be 25,500 students, the presumed high-level manpower requirement for the economy of 1980.

5. Semi-Logarithmic Graphs

In XII we noted that logarithmic graphs represent *proportionate* changes. This fact is often of use in portraying economic growth rates.

Let us use a hypothetical example. Suppose there are two firms with the following profit record.

	1960	*1961*	*1962*	*1963*	*1964*
Firm A	20	40	80	100	150
Firm B	140	160	200	220	270

The *absolute* increase in profits is the same for both firms: 20 in the first year, 40 in the second, 20 in the third and 50 in the fourth. This fact would be clearly shown on a natural scale.

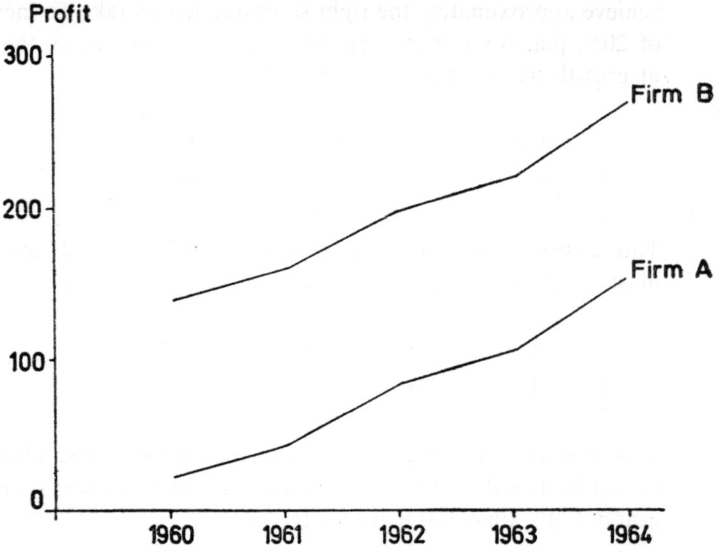

This information about the profit record of Firm A and Firm B is, however, much less interesting than the fact that A is growing very rapidly—much more rapidly than B. For example, in the first year A's profits doubled while B's increased only by 14%. This information could be shown on a semi-logarithmic graph. If the profits of the two firms are translated into logarithms we have

	1960	*1961*	*1962*	*1963*	*1964*
Firm A	1.3030	1·6021	1·9031	2·0000	2·1761
Firm B	2.1461	2·2041	2·3010	2·3424	2·4314

The resulting graph is much more revealing about the comparative rates of growth.

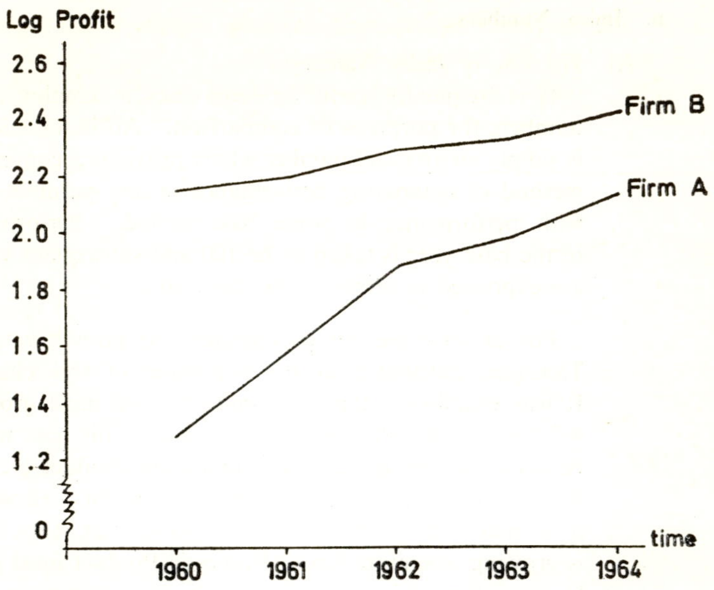

Another method of drawing a semi-logarithmic graph is to use a logarithmic scale on the vertical axis. This scale will show equal intervals for equal proportionate changes. Thus the spaces between 25 and 50, 50 and 100, 100 and 200 will be equal because each interval represents an increase of 100%. If this method is adopted, there is no need to translate the original figures into logarithms.

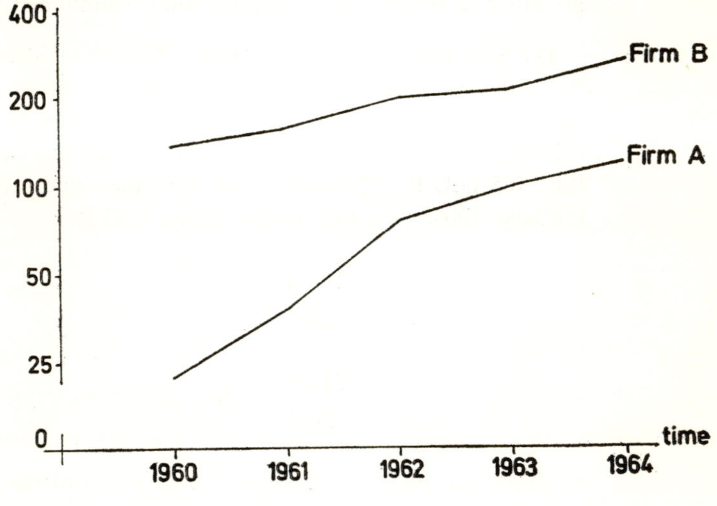

This graph gives exactly the same result as the preceding one. Both interpret the information in terms of comparative rates of growth.

6. Index Numbers

(i) *The Use of Index Numbers*

It is frequently useful in economics to employ index numbers for purposes of comparison. An index number is simply an artificial number which provides a convenient method of comparing performance at any point in time with performance in some base period. Performance in the base year is taken to be 100 and subsequent values are expressed in terms of the base value.

For an example, let us consider the growth rate for Tanzania, calculated at the beginning of this chapter. It was established that *per capita* income must grow at 4·7% p.a. if the Plan is to be fulfilled. This goal would be accomplished if the Gross Domestic Product grew at 4·7% and population remained constant. However, it is much more plausible to suppose that population is growing: therefore Gross Domestic Product must grow by more than 4·7% if the *per capita* income is to rise by the required amount.

Suppose that in actual fact the Gross Domestic Product of Tanzania grew from £200 million in 1964 (the year before the Plan) to £214·4 million in 1965 and £229·4 million in 1966. Over the same period population grew from 9·5 million people to 9·7 million to 9·9 million. Is the Plan being fulfilled? In other words, is the 4·7% growth rate of *per capita* income being maintained?

The base year in this case is 1964. Therefore to express the different values of the Gross Domestic Product as index numbers we give the value of the 1964 G.D.P. (£200m.) an index number of 100. The index number for the 1965 G.D.P. (£214.4m). must have the same ratio to the base 100 as the two values of the G.D.P.

$$\frac{x}{100} = \frac{214 \cdot 4}{200}$$

$$x = \frac{214 \cdot 4}{200} \cdot 100 = 107 \cdot 2$$

Similarly the 1966 G.D.P. expressed in terms of the base year 1964 is given by

$$\frac{229 \cdot 4}{200} \cdot 100 = 114 \cdot 7$$

Gross Domestic Product has risen from an index number of 100 to 107·2 in 1965 to 114·7 in 1966. From this we can readily ascertain that it rose 7·2% during the first year of the plan and 14·7 during the first two years of the plan. (Note: it did *not* rise 7·5% in the second year, because we are comparing both years to the Gross Domestic Product of 1964. To find the percentage increase in the second year it would be necessary to divide 229.4 by 214.4 which would show that it rose only 7·0% in the second year.)

To establish index numbers for the growth in population, again using 1964 as the base we have

$$\text{Pop. } 1965 = \frac{9 \cdot 7}{9 \cdot 5} \;.\; 100 = 102 \cdot 1$$

$$\text{Pop. } 1966 = \frac{9 \cdot 9}{9 \cdot 5} \;.\; 100 = 104 \cdot 2.$$

Has *per capita* income risen? Using index numbers, 1965 Gross Domestic Product is 107·2; 1965 population is 102·1 The growth in *per capita* income is

$$\frac{107 \cdot 2}{102 \cdot 1} \;.\; 100 = 104 \cdot 9.$$

In other words a 4.9% growth rate is being achieved: the plan is being fulfilled.

For the two year period the plan implies a growth rate of 9·6%: $(1 \cdot 047)^2 = 1 \cdot 0962$. Our index numbers show an increase in *per capita* income of

$$\frac{114 \cdot 7}{104 \cdot 2} \;.\; 100 = 110,$$

a growth rate of 10% for the two years; Tanzania is doing well.

Note. An approximate method frequently used by economists to find the rate of growth of *per capita* income is simply to subtract the rate of growth of population from the rate of growth of Gross Domestic Product. This method would show an increase of 5.1% for the first year, and an increase of 10.5% for the first two years.

(ii) *A Weighted Index*

An index frequently encountered is a weighted price index—a single index number, or series of index numbers, which measures the general level of prices at different periods in time. Let us suppose that in the two year period 1964—1966 consumer prices have risen in Tanzania, and it is feared that the apparent rise in the Gross Domestic Product can be partly accounted for by the higher price level. By how much have prices risen? Suppose surveys show that the 1966 prices of a typical consumer's weekly outlay are, in terms of a 1964 index of 100

Food, drink and tobacco	104
Clothing	101·2
Other items	101.

How can we find a single figure—the price index—which would indicate how prices in general have risen? A single price index *cannot* be calculated simply by averaging the three known numbers. The reason for this, as any wage-earner knows, is that in a typical week the consumer will spend far more on the first category than on the other two. Hence, a rise in the prices of food, drink and tobacco is much more important to him than a rise in other prices. Suppose that the average urban worker spends his weekly income as follows:

80% on food, drink and tobacco
15% on clothing
5% on other items.

A good price index will *weight* the various price changes in accordance with their importance in the consumer's budget. Each separate price index will be multiplied by its weight (80%, 15% and 5%); these totals are then combined and divided by the total number of weights (100%) to derive a *weighted average*. If P_1, P_2 and P_3 represent the three price changes, and W_1, W_2 and W_3 represent the weights, the price index is given by

$$I = \frac{P_1W_1 + P_2W_2 + P_3W_3}{W_1 + W_2 + W_3}$$

$$= \frac{104(\cdot 80) + 101\cdot 2(\cdot 15) + 101(.05)}{1\cdot 00}$$

$$= \frac{83\cdot 20 + 15\cdot 18 + 5\cdot 05}{1}$$

$$= 103\cdot 43$$

We are now in a position to reassess the apparent growth in *per capita* G.D.P. between 1964 and 1966. We found that the 1966 *per capita* G.D.P. index was 110, but now this figure must be "deflated" by the price index.

$$\frac{110}{103 \cdot 43} \, . \, 100 = 106{\cdot}3.$$

This shows that the rate of growth of *real per capita* income has been 6·3%—not the 7·6% rate called for by the plan.

In general, although the hypothetical Gross Domestic Product rose from £200m. to £229·4m. in the first two years of the development plan—in index numbers a rise from 100 to 114·7, implying a growth rate of 14·7%—this was not the whole story. Allowing for growth of population and price changes, real *per capita* income rose by only 6·3%.

EXERCISE 14

1. The population of Uganda in 1948 was approximately 4,918,000. In 1959 it was approximately 6,450,000.
   (a) What was the net increase in population?
   (b) What was the % increase in population?
   (c) What annual rate of growth does this increase imply?

2. The 1962 *per capita* income for Tanganyika was £20. Present development plans aim to raise *per capita* income to £45 by 1980. At what rate must the real Gross Domestic Product grow
   (a) if population is assumed to remain constant?
   (b) if population grows at 2·2% p.a.?

3. Calculate the rate of growth of *per capita* income in Tanganyika from £17·3 in 1957 to £20 in 1962, and compare this result with the answer to question (2).

4. In how many years will population double if it grows
   (a) at 2% per annum?
   (b) at 2·5% per annum?
   (c) at 3·0% per annum?

5. The value of Uganda's total exports declined from £52,157,000 in 1957 to £47,777,000 in 1962.
   (a) What rate of decline does this imply?
   (b) What possible explanations can be given for a decline in value?

6. The Economic Research Institute of the U.S.S.R. reckons that in 1958, Soviet industrial output equalled about 60% of the United States industrial ouput of that year. If the annual Soviet rate of growth is 7·4%, and the U.S. rate of growth is 3·2%, in which year will the industrial output of the two countries be equal?

7. The following figures show how the exports of two of Tanganyika's crops increased between 1958 and 1961.

	1958	*1959*	*1960*	*1961*
Tea (tons)	2,339	2,687	3,160	3,163
Cashew nuts (tons)	31,310	33,209	36,718	39,962

Express these two series in index numbers and determine which export trade grew faster.

8. By plotting the following information on graphs with
(a) a natural scale
(b) a semi-logarithmic scale,
determine which year saw the largest absolute increase and which year saw the largest proportionate increase, in the Gross Revenue of the East African Airways Corporation. Are these two answers the same?

Gross Revenue: E.A. Airways. (£'000)

1957	1,885	*1961*	4,967
1958	2,634	*1962*	6,412
1959	3,322	*1963*	7,623
1960	4,340		

9. Country A produced the following quantities of tractors, cloth and coffee in 1954 and 1964.

	1954	*1964*	*% value of total product*
Coffee ('000 tons)	600	550	60%
Cloth ('000 yards)	500	1250	10%
Tractors ('00s)	80	120	30%

Using 1954 as the base year, construct
(a) index numbers showing the output of coffee, cloth and tractors in 1964.
(b) A single index showing the rate of growth for the whole economy.

Does this overall production index seem to you to be a useful one?

10. *Nairobi Cost of Living* (1939 = 100)

	1960	*1962*	*Weights*
Food etc.	296	320	70
Clothing	272	276	15
Household	263	272	5
Servants	447	509	5
Other	210	230	5

(a) Construct a 1962 cost of living index using 1960 as the base year.

(b) Income between 1960 and 1962 increased from £225m. to £243m. Express the 1962 money income as an index number, 1960 = 100, and compare this with the answer to (a) to determine whether there was any change in *real* income, 1960—1962.

PART FOUR
SOME EXTENSIONS OF ELEMENTARY MATHEMATICS

XV.
PERMUTATIONS AND COMBINATIONS THE BINOMIAL THEOREM

1. Probability

The mathematics in this section is useful in the study of probability. The concept of probability is basic to economic reasoning because the complete certainty of a mathematical or logical argument is not possible when we come to deal with the real world. To take one example, price theory assumes that under normal conditions the quantity offered for sale in a market depends upon the price: the higher the price the greater the quantity, the lower the price the smaller the quantity. Is this assumption valid? Is there any evidence that when the price paid to coffee-growers declines over a period of years, farmers cut back their production of coffee? The answer to this question cannot be given with certainty. However, surveys may show that such text-book assumptions have at least some relevance to the real world. While it is not certain that coffee farmers will react predictably it may at least be *probable*. The problem then is to measure the degree of probability. Is a given event highly probable, quite likely or just possible?

Some probabilities cannot be measured with any precision. It would be difficult to measure the probability that the population of East Africa will continue to grow at an approximate rate of 3% per annum as the general standard of living rises. (Nevertheless, economic planners must make some assumption if their work is to have any meaning.) Other probabilities can be measured with a high degree of accuracy; this is where statistical methods come in.

The concept of probability is very difficult to define. Most formal definitions prove on inspection to be circular; i.e. an event is said to be probable if it is likely to happen. (What do we mean by "likely"?) For this reason, statisticians define probability in terms of relative frequency. A series of trials or observations is made and the number of occurrences of the event (E) is noted. Probability is then defined as the ratio of occurrences of E to the total number of observations. As the number of observations increases so the reliability of the estimate is presumed to increase

$$\Pr(E) = \lim_{n \to \infty} \frac{x}{n}$$

where n represents number of observations and x represents the number of occurrences of E.

If, for example, a survey of hospitals showed that out of 1000 live births 526 were boys and 474 were girls then the probability of a child being a boy would be defined

$$\Pr(B) = \frac{526}{1000} = \cdot 526.$$

Probability will always be between 0 and 1. Zero represents complete impossibility—the event cannot happen; one represents absolute certainty.

Two fundamental rules of probability may be noted here:

I. If two events E and F are mutually exclusive (cannot happen together), then the probability of either E or F is given by the sum of their separate probabilities:

$$\Pr(E \text{ or } F) = \Pr(E) + \Pr(F).$$

Example

There are ten equally suitable applicants for a job. Four are Kenyans, five are Ugandans, and one is a Tanzanian. If we assume the selection committee has no national prejudices, what is the probability of either a Kenyan or a Tanzanian will be chosen?

Probability of a Kenyan $= 4/10 = \cdot 4$

Probability of a Tanzanian $= 1/10 = \cdot 1$

$\Pr(K \text{ or } T) = \cdot 4 + \cdot 1 = \cdot 5$

II. If two events E and F are independent of each other then the probability that both events will occur is the product of their separate probabilities:

$$\Pr(E \text{ and } F) = \Pr(E).\ \Pr(F)$$

Example

If the probability that A will be alive in 20 years is 0·5, and the probability that B will be alive in 20 years is 0·7, what is the probability that both will be alive 20 years from now?

$$\Pr(A) = 0{\cdot}5$$
$$\Pr(B) = 0{\cdot}7$$
$$\Pr(A \text{ and } B) = (0{\cdot}5)\,(0{\cdot}7) = 0{\cdot}35.$$

The following paragraphs deal with particular aspects of mathematical probability: permutations, combinations, and the Binomial Theorem,useful for calculating probabilities in situations where events may happen in either of two mutually exclusive ways.

2. Terminology: Combinations, Permutation and "Factorial n"

(i) *Combinations and Permutations*

Suppose there are four objects denoted by A, B, C and D, and we wish to select groups of two. There are six possible selections:

AB, AC, AD, BC, BD, CD.

Each group of two is known as a *combination.*

If, however, we are also concerned with the order of letters in each group of two, there are twelve possibilities:

AB, AC, AD, BC, BD, CD
BA, CA, DA, CB, DB, DC.

Each arrangement is known as a *permutation.*

In forming combinations we are interested only in choosing a number of objects (in this case two letters); in forming permutations we are interested both in choosing a number of objects and in arranging them in order. For example: any combination of three students could be chosen to represent undergraduate opinion at a conference but if they are to be classified in order of merit as orators then the various permutations must also be considered. (There are six different ways of ordering a group of three people: 123, 132, 213, 231, 312, 321.)

(ii) *Factorial n*

Factorial n, denoted by n! is defined as

$$n! = n(n - 1)(n - 2) \dots (2)(1).$$

$$\text{Thus } 6! = 6 \cdot 5 . 4 . 3 . 2 . 1 = 720.$$

(Note. For convenience we define $0! = 1$)

3. Permutations

(i) A general expression for the number of permutations of n objects taken r at a time is given by

$$nPr = n(n-1)(n-2) \dots (n-r+1)$$

$$= \frac{n!}{(n - r)!}$$

The formula is derived in the following way. If we have n objects to be taken r at a time we may think of each member of a group of r being placed first, second, third, etc. The first place can be filled in n different ways, for we have n possibilities at our disposal. The second place may be filled in (n − 1) ways for there are only (n − 1) possibilities available to fill it, once the first place has been filled. Each way of filling the first place may be associated with each way of filling the second so that the first two places may be filled in n(n − 1) ways. Proceeding in this way, the first three places may be filled in n(n − 1) (n − 2) ways and all r places may be filled in

$$n(n-1)(n-2)\ldots.(n-r+1)$$

ways. That this expression is equivalent to

$$\frac{n!}{(n-r)!}$$

may be seen most readily from an example.

Example

A newly-formed committee has met to elect its officers. They wish to appoint a Chairman, a Vice-chairman, a Secretary and a Treasurer. If there are ten candidates, in how many ways can the four positions be filled? There are 10 possible ways of filling the first post, then 9 possible ways of filling the second, 8 possible ways of filling the third, and 7 ways of filling the fourth.

Therefore, there are
10.9.8.7 = 5040 possible ways.
Using the formula

$$_nP_r = n(n-1)\ldots..(n-r+1)$$
$$_{10}P_4 = 10.9.8.7 = 5040.$$

This is equivalent to

$$_nP_r = \frac{n!}{(n-r)!}$$

$$_{10}P_4 = \frac{10!}{6!} = \frac{10.9.8.7.6.5.4.3.2.1}{6.5.4.3.2.1} = 5040$$

(ii) A particular case of the general expression $_nP_r$ is given by $_nP_n$, that is the number of permutations of n objects taken n at a time.

$$_nP_n = n(n-1)(n-2)\ldots(1) = n!$$

Example

In how many ways can the first 6 letters of the alphabet be arranged?

$_6P_6 = 6.5.4.3.2.1 = 720.$

(iii) The number of permutations of n objects taken n at a time, when the group contains n_1 objects which are exactly alike, n_2 objects which are exactly alike, n_3 objects .... is given by

$$\frac{n!}{n_1!\, n_2!\, n_3! \ldots .} \quad \text{where } n = n_1 + n_2 + n_3 \ldots$$

Example

In how many ways can the letters of the word MAKERERE be arranged? There are three E's and two R's so the number of ways is

$$\frac{n!}{n_1!\, n_2!} = \frac{8!}{3!\, 2!} = 3360.$$

4. Combinations

(i) A general expression for the number of combinations of n objects taken r at a time is

$$_nC_r = \frac{_nP_r}{r!} = \frac{n(n-1) \ldots (n-r+1)}{r!} = \frac{n!}{r!(n-r)!}$$

The formula is derived in the following way The total number of arrangements of n objects taken r at a time is given by

$$_nP_r = n(n-1)\,(n-2) \ldots .\ (n-r+1)$$

In forming combinations, however, we are only interested in the possible number of groups of r and not in the order of the component objects. Each group of r objects may be arranged (see 3 ii) in r! different ways. Therefore, to find the number of combinations the total number of permutations must be divided through by r!

$$_nP_r = \frac{_nP_r}{r!}$$

This may be expressed in either of the two forms given above.

Example

An examination paper has twelve questions and candidates are expected to answer four. How many combinations of four questions are there?

$$_nC_r = \frac{n(n-1)(n-2)\ldots(n-r+1)}{r!}$$

$$_{12}C_4 = \frac{12.11.10.9}{4.3.2.1.} = 495.$$

(ii) In the above example it is obvious that for each combination of four questions the candidate chooses to answer, there must be eight that he chooses not to answer. Hence the number of combinations of four questions chosen must be exactly equal to the number of combinations of eight questions not chosen

$$_{12}C_4 = {}_{12}C_8$$

or more generally

$$_nC_r = {}_nC_{n-r}.$$

It should also be noted that there is only one way of choosing a group of n objects from n which is to take them all

$$_nC_n = 1.$$

5. The Binomial Theorem

(i) *Introduction*

The binomial theorem shows how to expand

$$(a + b)^n$$

where a and b are constants and n represents the positive integers (1, 2, 3, 4. . . .). Each term in the expansion will represent some combination of powers of a and b, and the rules for combinations form the basis for the theorem.

The binomial theorem has a wide application in statistics as it may be used to estimate probabilities in any case where there are only two mutually exclusive alternatives. For example, a sample of a consignment of fruit for canning may be tested, and each item will either be up to the required standard or it will not. Let the probability of any one item being up to standard be denoted by "a", and the probability of its being below standard be denoted by "b". If only two items are tested then the probability of both passing (by the second rule of probability given in 1.) is a^2. The probability of the first passing, and the second failing is ab; the probability of the first failing but the second passing is ba; and the probability that both will fail to reach the required standard is b^2. Summarizing these results we have

Two passes	One pass, one fail	Two fails
a^2	$2ab$	b^2

This is easy to recognize as the expansion of $(a+b)^2$.

If three items were tested then the possible results may be calculated in the same way, and it will be found that the probability of all three items being up to standard is a^3; the probability of only two being up to standard is $3a^2b$; the probability of only one being up to standard is $3ab^2$; and the probability of all being sub-standard is b^3. This set of results gives us the expansion of $(a + b)^3$.

Example

If it is feared that a consignment of electric light bulbs contains 5% defectives, what is the probability that a sample of three will reveal two good bulbs and one defective?

Let a = probability of good bulb = ·95

b = probability of defective bulb = ·05

Probability of two good and one defective bulb is

$$3a^2b = 3 \times \cdot 95 \times \cdot 95 \times \cdot 05 = \cdot 1354.$$

In general, for a sample of n items we need the expansion of $(a + b)^n$.

(ii) *The Theorem*

By multiplication it can be shown that

$$(a + b)^1 = a + b$$
$$(a + b)^2 = a^2 + 2ab + b^2$$
$$(a + b)^3 = a^3 + 3a^2b + 3ab^2 + b^3$$
$$(a + b)^4 = a^4 + 4a^3b + 6a^2b^2 + 4ab^3 + b^4$$
$$(a + b)^5 = a^5 + 5a^4b + 10a^3b^2 + 10a^2b^3 + 5ab^4 + b^5$$

etc.

Observation of the above table brings out two points. In the first place, the *exponent* of "a" decreases as we move term by term from left to right: the exponent of the first term being the same as the power to which (a + b) is raised

$a^n, a^{n-1}, a^{n-2} \dots a^2, a^1, a^0;$

the exponent of "b" increases from left to right, the exponent of the last term being the same as the power to which (a + b) is raised

$b^0, b^1, b^2, \dots b^{n-2}, b^{n-1}, b^n.$

In the second place, the numerical *coefficients of* the terms behave in the following way: the coefficient of the first and last term is always unity; the coefficients of the middle terms are:

in $(a + b)^2$ $_2C_1$

in $(a + b)^3$ $_3C_1, {}_3C_2$

in $(a + b)^4$ $_4C_1, {}_4C_2, {}_4C_3$

in $(a + b)^5$ $_5C_1, {}_5C_2, {}_5C_3, {}_5C_2$.

This suggests that in any expansion of $(a + b)^n$ the coefficients will be as follows:

first term	1
second ,,	$_nC_1$
third ,,	$_nC_2$
fourth ,,	$_nC_3$
⋮	
r^{th}	$_nC_{r-1}$

Example
Find the coefficient of the 4th term in the expansion of $(a + b)^8$. Let the 4th term equal the rth term, then $r - 1 = 3$, and the wanted coefficient

$$_nC_r = {}_8C_3 = \frac{8.7.6}{3.2} = 56.$$

Combining these results we have:

The binomial theorem for a positive integral index

$$(a + b)^n = a^n + {}_nC_1\, a^{n-1}b + {}_nC_2\, a^{n-2}b^2 + \ldots\ldots {}_nC_r\, a^{n-r}b^r \ldots\ldots + {}_nC_{n-1}\, ab^{n-1} + b^n.$$

Example

(a) Expand $(1 + x)^7$

$${}_7C_1 = 7$$

$${}_7C_2 = \frac{7.6}{2} = 21$$

$${}_7C_3 = \frac{7 \cdot 6 \cdot 5}{3 \cdot 2} = 35$$

and so the expansion begins
$1 + 7x + 21x^2 + 35x^3 \ldots.$

Since ${}_nC_r = {}_nC_{n-r}$

$${}_7C_4 = {}_7C_3,\ {}_7C_5 = {}_7C_2,\ {}_7C_6 = {}_7C_1$$

and the full expansion is
$1 + 7x + 21x^2 + 35x^3 + 35x^4 + 21x^5 + 7x^6 + x^7.$

(b) Expand $(1 - x)^7$

In this case all terms containing odd powers of x (x^1, x^3 ....) will be negative and the expansion becomes
$1 - 7x + 21x^2 - 35x^3 + 35x^4 - 21x^5 + 7x^6 - x^7.$

6. Pascal's Triangle

If the coefficients in the expansions $(a+b)$, $(a+b)^2$, $(a+b)^3$ .... are written in rows below one another we have:

Coefficients of $(a + b)^n$

n = 1						1		1					
n = 2					1		2		1				
n = 3				1		3		3		1			
n = 4			1		4		6		4		1		
n = 5		1		5		10		10		5		1	
n = 6	1		6		15		20		15		6		1

It will be seen that each row in the pyramid can be derived from the one above it. Each figure represents the sum of the two figures immediately above it. Thus the coefficient of the third term in the expansion of $(a + b)^5$ is 10, the sum of 4 and 6. The next row of the pyramid begins 1, 7, 21 .... ; this result may be compared with the example in the preceding paragraph.

Pascal's triangle provides a useful short cut to the calculation of coefficients in the expansion of $(a + b)^n$ for small values of n.

EXERCISE 15.

1. Assume that there are 9 coloured balls in a box; 2 are red, 4 are white, and 3 are black. If the following selections are made at random, calculate the probability of

   (a) drawing a red ball
   (b) drawing *either* a red ball *or* a white ball
   (c) drawing on two successive draws first a red ball and secondly a black ball. (Assume the first ball is not replaced.)

2. Evaluate ${}_8P_3$, ${}_6P_4$, ${}_{15}P_1$, ${}_3P_3$.

3. In how many different ways can the letters of

   (a) KENYA
   (b) UGANDA
   (c) TANZANIA
   be arranged?

4. How many 4 digit numbers can be formed with the 10 digits 0, 1, 2, 3, ... 9, if

   (a) repetitions are not allowed?
   (b) repetitions are allowed?
   (NB. No number can start with 0.)

5. There are 7 candidates for the three positions of President of the Students' Guild, Vice-President and Secretary. In how many ways can the positions be filled?

6. In how many ways can 13 books be arranged on a shelf if there is only room for 5 of them?

7. Evaluate ${}_7C_4$, ${}_{10}C_6$, ${}_6C_5$, ${}_4C_4$.

8. In how many ways can a team of 11 be chosen from 15 people?

9. If 10 people want to play tennis and only one court is available, how many groups of 4 can be chosen?

10. In how many ways can a deputation of 5 be chosen from a group of 9? If one man has to be included in the deputation, how many possible combinations are there?

11. What is the fifth term in the expansion $(x + y)^8$?

12. What is the coefficient of the third term of $(a + b)^{13}$?

13. Expand

(a) $(1 + x)^5$
(b) $(1 - x)^5$

14. Expand $(x + 2y)^6$ and apply the expansion to evaluate $(1{\cdot}02)^6$ correct to 5 places of decimals.
(Suggestion: let $x = 1$, $y = 10^{-2}$)

15. Use the binomial theorem to evaluate $0.902^{2\cdot 2}$ correct to four significant figures. Check the result by using logarithms.

XVI. INTRODUCTION TO MATRICES

1. Introduction

A matrix is a rectangular array of numbers enclosed by a pair of brackets. A matrix may be formed from a number of different fields of study; some examples follow:

Examples

(a)

$$\begin{bmatrix} 1 & -3 & 7 \\ 4 & 1 & -5 \end{bmatrix}$$

is a matrix formed from the coefficients of the two equations

$$x - 3y + 7z = 0$$
$$4x + y - 5z = 0$$

(b)

$$\begin{bmatrix} 2 & 4 \\ 3 & 9 \\ 4 & 16 \end{bmatrix}$$

is a matrix formed from the co-ordinates of three points P(2,4), Q(3,9) and R(4,16).

(c)

$$\begin{bmatrix} 0 & 15 & 0 & 0 & 0 & 0 \\ 5 & 0 & 5 & 0 & 5 & 0 \\ 0 & 0 & 0 & 0 & 0 & 0 \\ 0 & 2 & 2 & 0 & 3 & 3 \\ 5 & 3 & 5 & 0 & 0 & 2 \\ 0 & 0 & 0 & 0 & 0 & 0 \end{bmatrix}$$

is a matrix formed from the inter-industry transactions section of a hypothetical input-output for East Africa. The original table (see *African Economic Problems*, Vol. 2, Book 1, p. 442) reads

Input provided by \ Output sold to	Primary Products	Manufacturing	Construction	Public Utilities	Transport & Communication	Trade
Primary Products	..	15	..	..	..	..
Manufacturing	5	..	5	..	5	..
Construction	..	..	..	..	..	..
Public Utilities	..	2	2	..	3	3
Transport and Communications	5	3	5	..	..	2
Trade	..	..	..	..	..	..

These three examples suggest some fields in which matrices may be useful. The economist is likely to encounter matrices when solving systems of linear equations representing for example a set of Demand conditions, or a number of constraints on production. Even more important, the whole field of input-output analysis depends upon matrix algebra. In this Outline a few of the simpler matrix operations are explained.

2. Matrix Notation

The general form of a matrix is

$$A = \begin{bmatrix} a_{11} & a_{12} & a_{13} & \dots & a_{1n} \\ a_{21} & a_{22} & a_{23} & \dots & a_{2n} \\ a_{31} & a_{32} & a_{33} & \dots & a_{3n} \\ \dots & \dots & \dots & \dots & \dots \\ a_{m1} & a_{m2} & a_{m3} & \dots & a_{mn} \end{bmatrix}$$

This matrix has m rows and n columns and may be described as a matrix of *order* "m by n" or "m $\times$ n".

Each number in the matrix is known as an *element* and is represented by an expression of the form

$$a_{ij}$$

The two subscripts i and j show the exact position of the number in the matrix. The first subscript (i) indicates the row in which the number is to be found; the second subscript (j) indicates the column. Thus, the first subscript of all the numbers in the second row is two: a_{21}, a_{22}, a_{23} ..... a_{2n}, and the second subscript of all the numbers in the third column is three: a_{13}, a_{23}, a_{33} .... a_{m3}. The number a_{31} is the number in the third row and the first column of the matrix A, and in general the element a_{ij} is to be found in the ith and the jth column.

The matrix may be generally described as the

$$m \times n \text{ matrix } A = [a_{ij}]$$

3. Some Further Definitions

(i) *Vectors*

If either the number of rows or the number of columns in a matrix is equal to one, then the matrix A is a vector. A vector is a special case of a matrix.

If the number of rows is equal to one the matrix (vector) is known as a row vector.

Example

$$\begin{bmatrix} a_{11} & a_{12} & a_{13} & a_{14} \end{bmatrix} = \begin{bmatrix} 18 & 0 & 7 & -2 \end{bmatrix}$$

If the number of columns is equal to one, the matrix (vector) is known as a column vector.

Example

$$\begin{bmatrix} a_{11} \\ a_{21} \\ a_{31} \end{bmatrix} = \begin{bmatrix} 1,437 \\ 20,215 \\ 5,672 \end{bmatrix}$$

(ii) *Square Matrices*

If the number of rows in a matrix is equal to the number of columns (m = n) then the matrix A is a square matrix.

Example

$$\begin{bmatrix} a_{11} & a_{12} & a_{13} \\ a_{21} & a_{22} & a_{23} \\ a_{31} & a_{32} & a_{33} \end{bmatrix} = \begin{bmatrix} 17.2 & 14.3 & 18.7 \\ 19.2 & 14.8 & 17.6 \\ 16.7 & 18.2 & 15.4 \end{bmatrix}$$

(iii) *Symmetric Matrices*

If the matrix A is unaffected by transposing the rows and columns (so that the first row becomes the first column; the second row becomes the second column etc.) the matrix is said to be symmetric.

Example

$$\begin{bmatrix} a_{11} & a_{12} & a_{13} \\ a_{21} & a_{22} & a_{23} \\ a_{31} & a_{32} & a_{33} \end{bmatrix} = \begin{bmatrix} 1 & 2 & 3 \\ 2 & 4 & -5 \\ 3 & -5 & 6 \end{bmatrix}$$

Another way of defining a symmetric matrix is to say that a matrix is symmetric if

$$a_{ij} = a_{ji}$$

for all elements of the matrix. For example the matrix above is symmetric because

$$a_{11} = a_{11} = 1$$

$$a_{12} = a_{21} = 2$$

$$a_{13} = a_{31} = 3$$

$$a_{22} = a_{22} = 4$$

$$a_{23} = a_{32} = -5$$

$$a_{33} = a_{33} = 6$$

(iv) *Sub-matrices*

A sub-matrix may be formed from any matrix A by striking out one or more rows or columns from the original matrix.

Example

$$\begin{bmatrix} -2 & 3 \\ 9 & 4 \end{bmatrix}$$ is a sub-matrix of

$$\begin{bmatrix} -2 & 3 & 8 & 1 \\ 9 & 4 & 6 & 0 \\ 1 & 15 & 2 & -7 \end{bmatrix}$$

(v) *The Identity Matrix*

If each number in the diagonal of a square matrix a_{11} a_{22} a_{33} . . . a_{nn} is equal to one; and all the other numbers are equal to zero, the matrix A is said to be an identity matrix. The identity matrix may be denoted by I_n (where n shows the order of the matrix).

Example

$$I = \begin{bmatrix} a_{11} & a_{12} & a_{13} & a_{14} \\ a_{21} & a_{22} & a_{23} & a_{24} \\ a_{31} & a_{32} & a_{33} & a_{34} \\ a_{41} & a_{42} & a_{43} & a_{44} \end{bmatrix} = \begin{bmatrix} 1 & 0 & 0 & 0 \\ 0 & 1 & 0 & 0 \\ 0 & 0 & 1 & 0 \\ 0 & 0 & 0 & 1 \end{bmatrix}$$

(vi) *The Transpose of a Matrix*

If the rows and columns of the matrix A are interchanged (so that the first row becomes the first column, the second row becomes the second column etc.) the new matrix is known as the transpose of A.

Example

$$\text{If } A = \begin{bmatrix} a_{11} & a_{12} & a_{13} \\ a_{21} & a_{22} & a_{23} \end{bmatrix} = \begin{bmatrix} 1 & 2 & 3 \\ 4 & 5 & 6 \end{bmatrix}$$

$$A \text{ transpose} = \begin{bmatrix} a_{11} & a_{21} \\ a_{12} & a_{22} \\ a_{13} & a_{23} \end{bmatrix} = \begin{bmatrix} 1 & 4 \\ 2 & 5 \\ 3 & 6 \end{bmatrix}$$

Notice that the element a_{ij} stands in the jth row and the ith column of the transpose of the matrix A, while in the original matrix A, the element a_{ij} stands in the ith row and the jth column.

5. Simple Operations with Matrices

(i) *Addition and Subtraction*

Addition or subtraction of matrices consists of adding or subtracting the corresponding elements in each matrix. Two matrices are said to be *conformable* for addition or subtraction if they are both of the same order $m \times n$.

$$\text{If } A = \begin{bmatrix} a_{11} & a_{12} & \dots & a_{1n} \\ a_{21} & a_{22} & \dots & a_{2n} \\ \dots & \dots & \dots & \dots \\ a_{m1} & a_{m2} & \dots & a_{mn} \end{bmatrix}$$

$$\text{and } B = \begin{bmatrix} b_{11} & b_{12} & \dots & b_{1n} \\ b_{21} & b_{22} & \dots & b_{2n} \\ \dots & \dots & \dots & \dots \\ b_{m1} & b_{m2} & \dots & b_{mn} \end{bmatrix}$$

$$A + B = \begin{bmatrix} a_{11} + b_{11} & a_{12} + b_{12} & \dots & b_{1n} + b_{1n} \\ a_{21} + b_{21} & a_{22} + b_{22} & \dots & a_{2n} + b_{2n} \\ \dots & \dots & \dots & \dots \\ a_{m1} + b_{m1} & a_{m2} + b_{m2} & \dots & a_{mn} + b_{mn} \end{bmatrix}$$

$$A - B = \begin{bmatrix} a_{11} - b_{22} & a_{12} - b_{12} & \dots & a_{1n} - b_{1n} \\ a_{21} - b_{21} & a_{22} - b_{22} & \dots & a_{2n} - b_{2n} \\ \dots & \dots & \dots & \dots \\ a_{m1} - b_{m1} & a_{m2} - b_{m2} & \dots & a_{mn} - b_{mn} \end{bmatrix}$$

Examples

(a)

$$\begin{bmatrix} 1 & 2 & 3 \\ 0 & 1 & 4 \end{bmatrix} + \begin{bmatrix} 2 & 3 & 0 \\ -1 & 2 & 5 \end{bmatrix} = \begin{bmatrix} 3 & 5 & 3 \\ -1 & 3 & 9 \end{bmatrix}$$

(b)

$$\begin{bmatrix} 5 & 0 & 7 \\ -2 & 1 & 12 \\ 5 & -6 & 1 \end{bmatrix} - \begin{bmatrix} 3 & 1 & 1 \\ 8 & 11 & -5 \\ -4 & 6 & -3 \end{bmatrix} = \begin{bmatrix} 2 & -1 & 6 \\ -10 & -10 & 17 \\ 9 & -12 & 4 \end{bmatrix}$$

(ii) *Multiplication of Matrices*

If the matrix A is a row vector of order $1 \times m$

$$A = \begin{bmatrix} a_{11} & a_{12} & a_{13} & \cdots & a_{1m} \end{bmatrix}$$

and the matrix B is a column vector of order $m \times 1$

$$B = \begin{bmatrix} b_{11} \\ b_{21} \\ b_{31} \\ \cdot \\ \cdot \\ \cdot \\ b_{m1} \end{bmatrix}$$

then the product AB, in that order, is given by

$$AB = \begin{bmatrix} a_{11}b_{11} + a_{12}\ b_{21} + a_{13}b_{31} \ \cdots \ a_{1m}b_{m1} \end{bmatrix}$$

Each element in the row is multiplied into the corresponding element in the column. A common notation for this is

$$AB = \begin{bmatrix} \sum_{k=1}^{m} a_{1k}b_{k1} \end{bmatrix}$$

This simply means that the product AB is the sum (represented by Σ, the Greek letter sigma) of the products $a_{1k}b_{k1}$ for all values of k from 1 to m.

Example

$$\begin{bmatrix} 5 & 3 & 2 \end{bmatrix} \begin{bmatrix} 6 \\ -4 \\ 1 \end{bmatrix} = \begin{bmatrix} 5(6)+3(-4)+2(1) \end{bmatrix}$$

$$= 20.$$

Two matrices A and B are said to be conformable for multiplication if the number of columns in A is equal to the number of rows in B. In the above example the first vector has three columns and the second has three rows.

To multiply two matrices which are not vectors, the same principle of multiplying columns by rows is followed. If the matrix A is of order $m \times p$ and the matrix B is of order $p \times n$ then the product AB, in that order, is found by multiplying each row of A once and once only into each column of B. The resulting matrix C will be of order $m \times n$. This may be formally expressed as follows:

The product AB in that order of the $m \times p$ matrix $A = [a_{ij}]$ and the $p \times n$ matrix $B = [b_{ij}]$ is equal to the $m \times n$ matrix $C = [c_{ij}]$ where

$$
\begin{aligned}
C_{ij} &= a_{i1}b_{1j} + a_{i2}b_{2j} + \ldots\ldots a_{ip}b_{pj} \\
&= \sum_{k=1}^{p} a_{ik}b_{kj}
\end{aligned}
$$

$$(i = 1, 2, 3 \ldots . m;\ j = 1, 2, 3 \ldots .n)$$

Examples

(a)

$$
AB = \begin{bmatrix} a_{11} & a_{12} \\ a_{21} & a_{22} \\ a_{31} & a_{32} \end{bmatrix} \begin{bmatrix} b_{11} & b_{12} \\ b_{21} & b_{22} \end{bmatrix}
$$

$$
= \begin{bmatrix} a_{11}b_{11} + a_{12}b_{21} & a_{11}b_{12} + a_{12}b_{22} \\ a_{21}b_{11} + a_{22}b_{21} & a_{21}b_{12} + a_{22}b_{22} \\ a_{31}b_{11} + a_{32}b_{21} & a_{31}b_{12} + a_{32}b_{22} \end{bmatrix}
$$

In this example A is of order 3×2; B is of order 2×2; and the resulting matrix is of order 3×2.

(b)

$$
\begin{bmatrix} 1 & 2 & 1 \\ 4 & 0 & 2 \end{bmatrix} \begin{bmatrix} 3 & -4 \\ 1 & 5 \\ -2 & 2 \end{bmatrix}
$$

$$
= \begin{bmatrix} 1(3)+2(1)+1(-2) & 1(-4)+2(5)+1(2) \\ 4(3)+0(1)+2(-2) & 4(-4)+0(5)+2(2) \end{bmatrix}
$$

$$
= \begin{bmatrix} 3 & 8 \\ 8 & -12 \end{bmatrix}
$$

In this example, a 2×3 matrix is multiplied into a 3×2 matrix to produce a 2×2 matrix.

(c)
$$\begin{bmatrix} 1 & 0 & 0 \\ 0 & 1 & 0 \\ 0 & 0 & 1 \end{bmatrix} \begin{bmatrix} 6 & 4 \\ 3 & 1 \\ 2 & 8 \end{bmatrix}$$

$$= \begin{bmatrix} 6+0+0 & 4+0+0 \\ 0+3+0 & 0+1+0 \\ 0+0+2 & 0+0+8 \end{bmatrix}$$

$$= \begin{bmatrix} 6 & 4 \\ 3 & 1 \\ 2 & 8 \end{bmatrix}$$

This example shows that multiplying a matrix by an identity matrix does not alter its value.

6. A Simple Application of Matrix Algebra

Let the matrix

$$Q = \begin{bmatrix} q_{11} & q_{12} & q_{13} & \cdots\cdots & q_{1n} \\ q_{21} & q_{22} & q_{23} & \cdots\cdots & q_{2n} \\ q_{31} & q_{32} & q_{33} & \cdots\cdots & q_{3n} \\ \cdots & \cdots & \cdots & \cdots\cdots & \cdots \\ q_{m1} & q_{m2} & q_{m3} & \cdots\cdots & q_{mn} \end{bmatrix}$$

be interpreted as follows. Each column of the matrix Q represents a particular firm in an industry. There are n firms in the industry. Each row of the matrix represents a different type of raw material purchased by the firms in the industry. There are m different types of raw material. The number q_{ij} represents the quantity of the ith type of raw material purchased by firm j. For example q_{34} is the quantity of the third type of raw material purchased by the fourth firm in the industry.

Suppose Q (O) is a matrix representing purchases by the different firms of the various raw materials for the month of October. Let Q (N) and Q (D) be the matrices representing November and December purchases. The total quantity of raw materials purchased during the fourth quarter of the year by the different firms may be represented by the matrix Q (IV) where

$$Q\ (IV) = Q\ (O) + Q\ (N) + Q\ (D).$$

Suppose V (III) and V (IV) are matrices representing the stocks or inventories of the various types of raw materials held by the different firms at the end of the third quarter and the fourth quarter of the year. Then the total use of the various types of raw materials during the last quarter of the year can be represented by the matrix U (IV) where

$$U\ (IV) = V\ (III) + Q\ (IV) - V\ (IV).$$

In other words, inventories at the end of the third quarter plus purchases during the fourth quarter less inventories at the end of the fourth quarter, equals use during the fourth quarter.

To find the value of raw materials used by the firms in the fourth quarter it is necessary to know the prices of the different raw materials. Let the prices of the raw materials (1, 2, 3 . . . .m) be represented by $p_1, p_2, p_3 \ldots p_m$. The prices may be represented by a price vector

$$p = [p_{11}\ \ p_{12}\ \ p_{13}\ \ldots\ldots\ p_{1m}].$$

To calculate the total value of the raw materials used during the fourth quarter of the year the total use matrix U (IV) must be multiplied by the price vector p.

$$[p_{11}\ p_{12}\ p_{13} \ldots p_{1m}] \begin{bmatrix} u_{11} & u_{12} & u_{13} & \ldots & u_{1n} \\ u_{21} & u_{22} & u_{23} & \ldots & u_{2n} \\ u_{31} & u_{32} & u_{33} & \ldots & u_{3n} \\ \ldots & \ldots & \ldots & \ldots & \ldots \\ u_{m1} & u_{m2} & u_{m3} & \ldots & u_{mn} \end{bmatrix}$$

The product will be a $1 \times n$ vector showing the total value of the raw materials used by each firm in the industry during the fourth quarter of the year. This total value vector may be represented by

$$T = [t_{1j}]$$

$$\text{where } t_{1j} = \sum_{k=1}^{m} p_{1k}\, u_{kj} \quad (j = 1,2,3 \ldots . n).$$

Note. The subscript 1 is often omitted in vector notation. The above total value vector could be described simply as the matrix $T = [t_j]$.

EXERCISE 16

1. Which of the following pairs of matrices are conformable for addition? Multiplication? Neither?

(a) $\begin{bmatrix} 1 & 3 \\ 0 & -2 \end{bmatrix}$ and $\begin{bmatrix} 5 & 7 & 9 \\ -3 & 2 & 1 \end{bmatrix}$

(b) $\begin{bmatrix} 5 \\ 2 \\ -7 \end{bmatrix}$ and $\begin{bmatrix} 5 & 8 & 1 \end{bmatrix}$

(c) $\begin{bmatrix} 2 & 1 & 6 \\ 5 & -4 & -2 \end{bmatrix}$ and $\begin{bmatrix} 0 & 2 & 7 \\ -8 & 11 & 3 \end{bmatrix}$

(d) $\begin{bmatrix} 5 & 1 \\ -8 & 6 \\ 12 & 4 \end{bmatrix}$ and $\begin{bmatrix} 13 & -22 & 8 \\ 6 & 0 & 70 \end{bmatrix}$

(e) $\begin{bmatrix} 5 & 2 & 1 & 10 \end{bmatrix}$ and $\begin{bmatrix} 4 & -13 \\ -2 & 0 \\ 8 & 3 \\ 7 & 1 \end{bmatrix}$

2. Add:

(a) $\begin{bmatrix} 5 & 9 \\ -4 & -3 \end{bmatrix} + \begin{bmatrix} -2 & 1 \\ 6 & 5 \end{bmatrix}$

(b) $\begin{bmatrix} -3 & 1 & 6 \\ 2 & -4 & 3 \\ 4 & -2 & 0 \end{bmatrix} + \begin{bmatrix} 10 & 6 & -20 \\ 7 & 1 & 0 \\ 18 & 5 & -9 \end{bmatrix}$

3. Subtract:

(a) $\begin{bmatrix} 1 & 2 & -3 \\ 6 & 4 & 12 \end{bmatrix} - \begin{bmatrix} 0 & -7 & 6 \\ -2 & 1 & 4 \end{bmatrix}$

(b) $\begin{bmatrix} 5 & 8 \\ 3 & -12 \\ -12 & 1 \end{bmatrix} - \begin{bmatrix} 2 & 7 \\ -8 & 1 \\ -3 & 8 \end{bmatrix}$

4. Show by multiplication that

(a) $$\begin{bmatrix} 2 & 3 & 4 \\ 1 & 5 & 6 \end{bmatrix} \begin{bmatrix} 1 \\ 2 \\ 3 \end{bmatrix} = \begin{bmatrix} 20 \\ 29 \end{bmatrix}$$

(b) If $A = \begin{bmatrix} 2 & -1 & 1 \\ 0 & 1 & 2 \\ 1 & 0 & 1 \end{bmatrix}$

$A^2 = \begin{bmatrix} 5 & -3 & 1 \\ 2 & 1 & 4 \\ 3 & -1 & 2 \end{bmatrix}$

(c) $$\begin{bmatrix} 1 & 2 & 3 \\ 1 & 3 & 3 \\ 1 & 2 & 4 \end{bmatrix} \begin{bmatrix} 6 & -2 & -3 \\ -1 & 1 & 0 \\ -1 & 0 & 1 \end{bmatrix} = \begin{bmatrix} 1 & 0 & 0 \\ 0 & 1 & 0 \\ 0 & 0 & 1 \end{bmatrix}$$

5. Multiply:

(a) $\begin{bmatrix} 4 & 5 & 6 \end{bmatrix} \begin{bmatrix} 2 \\ 3 \\ -1 \end{bmatrix}$

(b) $\begin{bmatrix} 2 \\ 3 \\ -1 \end{bmatrix} \begin{bmatrix} 4 & 5 & 6 \end{bmatrix}$

(c) $\begin{bmatrix} 1 & 2 & 1 \\ 4 & 0 & 2 \end{bmatrix} \begin{bmatrix} 3 & -4 \\ 1 & 5 \\ -2 & 2 \end{bmatrix}$

(d) $\begin{bmatrix} 1 & 2 & 3 \end{bmatrix} \quad \begin{bmatrix} 4 & -6 & 9 & 6 \\ 0 & -7 & 10 & 7 \\ 5 & 8 & -11 & -8 \end{bmatrix}$

XVII. MATRICES CONTINUED. DETERMINANTS AND LINEAR EQUATIONS

1. Introduction

A matrix is defined as a rectangular array of numbers, and a square matrix is a matrix in which the number of rows is equal to the number of columns. To each square matrix A there corresponds a single value known as its *determinant* $|A|$.

2. Types of Determinants

(i) *Second order determinants*

A determinant of the second order is defined as follows:

$$|A| = \begin{vmatrix} a_{11} & a_{12} \\ a_{21} & a_{22} \end{vmatrix} = a_{11}a_{22} - a_{12}a_{21}$$

The value of the determinant is given by the difference of the two cross-products.

Example

$$\begin{vmatrix} 4 & 5 \\ 1 & 3 \end{vmatrix} = 4(3) - 5(1) = 7$$

(ii) *Third order determinants*

A determinant of the third order is defined in terms of the second-order determinant:

$$|A| = \begin{vmatrix} a_{11} & a_{12} & a_{13} \\ a_{21} & a_{22} & a_{23} \\ a_{31} & a_{32} & a_{33} \end{vmatrix}$$

$$= a_{11}\begin{vmatrix} a_{22} & a_{23} \\ a_{32} & a_{33} \end{vmatrix} - a_{12}\begin{vmatrix} a_{21} & a_{23} \\ a_{31} & a_{33} \end{vmatrix} + a_{13}\begin{vmatrix} a_{21} & a_{22} \\ a_{31} & a_{32} \end{vmatrix}$$

Example

$$\begin{vmatrix} 1 & 5 & 3 \\ 2 & 0 & 5 \\ -4 & 1 & -2 \end{vmatrix}$$

$$= 1\begin{vmatrix} 0 & 5 \\ 1 & -2 \end{vmatrix} - 5\begin{vmatrix} 2 & 5 \\ -4 & -2 \end{vmatrix} + 3\begin{vmatrix} 2 & 0 \\ -4 & 1 \end{vmatrix}$$

$$= (0 - 5) - 5(-4 + 20) + 3(2 - 0)$$

$$= -79.$$

(iii) *Fourth order determinants*

A determinant of the fourth order is defined in terms of third order determinants:

$$|A| = \begin{vmatrix} a_{11} & a_{12} & a_{13} & a_{14} \\ a_{21} & a_{22} & a_{23} & a_{24} \\ a_{31} & a_{32} & a_{33} & a_{34} \\ a_{41} & a_{42} & a_{43} & a_{44} \end{vmatrix}$$

$$= a_{11}\begin{vmatrix} a_{22} & a_{23} & a_{24} \\ a_{32} & a_{33} & a_{34} \\ a_{42} & a_{43} & a_{44} \end{vmatrix} - a_{12}\begin{vmatrix} a_{21} & a_{23} & a_{24} \\ a_{31} & a_{33} & a_{34} \\ a_{41} & a_{43} & a_{44} \end{vmatrix}$$

$$+ a_{13}\begin{vmatrix} a_{21} & a_{22} & a_{24} \\ a_{31} & a_{32} & a_{34} \\ a_{41} & a_{42} & a_{44} \end{vmatrix} - a_{14}\begin{vmatrix} a_{21} & a_{22} & a_{23} \\ a_{31} & a_{32} & a_{33} \\ a_{41} & a_{42} & a_{43} \end{vmatrix}$$

(iv) *Determinants of the* n^{th} *order*

A determinant of the nth order is defined in terms of determinants of order $(n-1)$.

$$|A| = \begin{vmatrix} a_{11} & a_{12} & a_{13} & \cdots & a_{1n} \\ a_{21} & a_{22} & a_{23} & \cdots & a_{2n} \\ a_{21} & a_{31} & a_{33} & \cdots & a_{3n} \\ \cdots & \cdots & \cdots & \cdots & \cdots \\ a_{n1} & a_{n2} & a_{n3} & \cdots & a_{nn} \end{vmatrix}$$

$$= a_{11}\begin{vmatrix} a_{22} & a_{23} & \ldots & a_{2n} \\ a_{32} & a_{33} & \ldots & a_{3n} \\ \ldots & \ldots & \ldots & \ldots \\ a_{n2} & a_{n3} & \ldots & a_{nn} \end{vmatrix} - a_{12}\begin{vmatrix} a_{21} & a_{23} & \ldots & a_{2n} \\ a_{31} & a_{33} & \ldots & a_{3n} \\ \ldots & \ldots & \ldots & \ldots \\ a_{n1} & a_{n3} & \ldots & a_{nn} \end{vmatrix}$$

$$+ \ldots\ldots\ldots$$

The signs of the successive terms in the expansion will be alternately positive and negative until the last term is reached.

3. Minors and Cofactors

In all the above examples the expansion of the nth order determinant was by the first row. The terms in the first row a_{11}, a_{12}, $a_{13}, \ldots a_{1n}$ were multiplied by the determinant of order $(n - 1)$ formed by eliminating the row and column in which the element a_{1j} stood.

This result may be generalized. A determinant of the nth order may be expanded by the elements in any one row or column. If the successive elements are denoted by a_{ij} (where i or j is constant) the $(n - 1)^{th}$ order determinant is formed by eliminating the ith row and the jth column. This determinant of order $(n - 1)$ is known as the *minor* of a_{ij} and may be denoted by the term

$$|M_{ij}|$$

The successive terms in the expansion of the nth order determinant will have positive or negative signs. The sign of the minor $|M_{ij}|$ is formed by multiplying $|M_{ij}|$ by $(-1)^{i+j}$. For example, the minor of the element a_{13} will have a positive sign $(-1)^{1+3} = 1$, while the minor of the element a_{14} will have a negative sign $(-1)^{1+4} = -1$. Briefly, if $i+j$ is even, the sign will be positive; if $i+j$ is odd, the sign will be negative. The resulting signed minor is known as the *cofactor* of a_{ij}. A useful notation for the cofactor of the element a_{ij} is

$$\alpha_{ij} = (-1)^{i+j} \; |M_{ij}| .$$

This new notation for cofactors makes it possible to write down the expansion of an nth order determinant more simply.

$$|A| = a_{i1}\alpha_{i1} + a_{i2}\alpha_{i2} + \ldots + a_{in}\alpha_{in} = \sum_{k=1}^{n} a_{ik}\alpha_{ik}$$

(expanding by the ith row)

$$\text{or } |A| = a_{1j}\alpha_{1j} + a_{2j}\alpha_{2j} + \ldots + a_{nj}\alpha_{nj} = \sum_{k=1}^{n} a_{kj}\alpha_{kj}$$

(expanding by the jth column).

4. Elementary Properties of Determinants

The following properties are useful in the evaluation of determinants.

I. A determinant is unchanged in value if its rows and columns are transposed.

II. A determinant changes sign if two rows (or columns) are interchanged.

III. The value of a determinant is increased k-fold if each element in one row (or column) is multiplied by k. Conversely, if every element in a row (or column) contains a common factor k, then k may be factored from the determinant.

IV. A determinant is equal to zero if two rows (or columns) are identical.

V. The value of a determinant is not affected if one row (or column) or a multiple thereof is added to, or subtracted from, another row (or column).

Examples

(a) Evaluate

$$|A| = \begin{vmatrix} 26 & 2 & 5 \\ 13 & 7 & 1 \\ 52 & 4 & 20 \end{vmatrix}$$

Factor 13 from the first column (property III).

$$= 13 \begin{vmatrix} 2 & 2 & 5 \\ 1 & 7 & 1 \\ 4 & 4 & 20 \end{vmatrix}$$

Subtract the first column from the second (pro– perty V).

$$= 13 \begin{vmatrix} 2 & 0 & 5 \\ 1 & 6 & 1 \\ 4 & 0 & 20 \end{vmatrix}$$

Expand by the second column.

$$= (13)\ (6) \begin{vmatrix} 2 & 5 \\ 4 & 20 \end{vmatrix}$$

$$= 13\ (6)\ (40 - 20)$$

$$= 1560.$$

(b) Prove that

$$|A| = \begin{vmatrix} 19 & 17 & 15 \\ 9 & 8 & 7 \\ 1 & 1 & 1 \end{vmatrix} = 0$$

Subtract twice the second row from the first (property V).

$$|A| = \begin{vmatrix} 1 & 1 & 1 \\ 9 & 8 & 7 \\ 1 & 1 & 1 \end{vmatrix}$$

The first row and the third row are now identical, therefore (property IV)

$$|A| = 0$$

5. Further Characteristics of Matrices

(i) *Singular and non-singular matrices*

A square matrix is said to be *singular* if its determinant is equal to zero.

$$|A| = 0$$

A square matrix is *non singular* if its determinant is not equal to zero

$$|A| \neq 0$$

(ii) *The rank of a matrix*

The rank of a matrix A is the order of the largest non-singular square sub-matrix to be found in it.

Examples

(a)

$$A = \begin{bmatrix} 4 & 8 & 2 & 1 \\ 5 & 7 & 1 & 0 \\ 5 & 3 & 9 & 2 \end{bmatrix}$$

This matrix has four different square sub-matrices of order three:

$$\begin{bmatrix} 4 & 8 & 2 \\ 5 & 7 & 1 \\ 5 & 3 & 9 \end{bmatrix}$$ obtained by striking out the last column of A,

$$\begin{bmatrix} 4 & 8 & 1 \\ 5 & 7 & 0 \\ 5 & 3 & 2 \end{bmatrix}$$ obtained by striking out the third column of A,

$$\begin{bmatrix} 4 & 2 & 1 \\ 5 & 1 & 0 \\ 5 & 9 & 2 \end{bmatrix}$$ obtained by striking out the second column of A,

$$\begin{bmatrix} 8 & 2 & 1 \\ 7 & 1 & 0 \\ 3 & 9 & 2 \end{bmatrix}$$ obtained by striking out the first column of A.

To consider the value of only one of these sub-matrices; the value of the last is

$$8(2-0) - 2(14-0) + (63-3) = 48$$
$$\neq 0$$

Therefore the rank of the matrix A is three, the order of the largest square non-singular sub-matrix. There are no square sub-matrices with an order greater than three.

(b)

$$A = \begin{bmatrix} 3 & 1 & 2 & 1 \\ 1 & 1 & 1 & 0 \\ 11 & 7 & 9 & 2 \end{bmatrix}$$

As in the previous example the largest square sub-matrices; are of order three:

$$\begin{bmatrix} 3 & 1 & 2 \\ 1 & 1 & 1 \\ 11 & 7 & 9 \end{bmatrix}, \begin{bmatrix} 3 & 1 & 1 \\ 1 & 1 & 0 \\ 11 & 7 & 2 \end{bmatrix},$$

$$\begin{bmatrix} 3 & 2 & 1 \\ 1 & 1 & 0 \\ 11 & 9 & 2 \end{bmatrix} \text{ and } \begin{bmatrix} 1 & 2 & 1 \\ 1 & 1 & 0 \\ 7 & 9 & 2 \end{bmatrix}$$

The determinant of each of these 3×3 sub-matrices is equal to zero. (The reader may satisfy himself that this is so). The next largest square sub-matrices are of order 2; one of them, for example, is

$$\begin{bmatrix} 3 & 2 \\ 1 & 1 \end{bmatrix},$$ obtained by striking out the third row, and the second and fourth columns of A.

The determinant of this sub-matrix is not equal to zero.

$$\begin{vmatrix} 3 & 2 \\ 1 & 1 \end{vmatrix} = 3 - 2 = 1.$$

Thus the matrix A has at least one square sub-matrix of order 2, and the rank of the matrix A is 2.

(iii) *The Inverse of a Matrix*

If there are two square matrices A and B such that the product $A \times B$ (and the product $B \times A$) is equal to the identity matrix I_n, then B is called the inverse of A, and A is called the inverse of B. The inverse of the matrix A is usually denoted by A^{-1}.

Example

$$\begin{bmatrix} 1 & 2 & 3 \\ 1 & 3 & 3 \\ 1 & 2 & 4 \end{bmatrix} \begin{bmatrix} 6 & -2 & -3 \\ -1 & 1 & 0 \\ -1 & 0 & 1 \end{bmatrix} = \begin{bmatrix} 1 & 0 & 0 \\ 0 & 1 & 0 \\ 0 & 0 & 1 \end{bmatrix}$$

therefore each matrix is the inverse of the other.

The relationship of the matrix A to its inverse A^{-1} is in many ways similar to that of x and $\frac{1}{x}$ in algebra. However, the inverse of a matrix is *not* simply the reciprocal of A. The method of calculating the inverse of a matrix is complex; the rules are set out below and can be applied to 2 × 2 or 3 × 3 matrices. For higher order matrices the method is the same, but without a computer the calculations are rather tedious.

The steps for calculating the inverse of a matrix A are as follows.

Step 1.

From the matrix A, form a new matrix by replacing every element a_{ij} by its cofactor which we may call c_{ij}.

Step 2.

Transpose this matrix, writing the rows as columns and the columns as rows. This new matrix is known as the *adjoint* of A and may be defined

$$\text{Adj}A = [c_{ji}]$$

Step 3.

Calculate the value of the determinant $|A|$

Step 4.

The inverse matrix can now be derived. Each element in the inverse is calculated by dividing the corresponding element in the adjoint by the value of the determinant $|A|$.

Examples

(a) Calculate the inverse of the 2 × 2 matrix

$$A = \begin{bmatrix} 3 & 1 \\ 2 & 4 \end{bmatrix}$$

Step 1.

$$\begin{bmatrix} 4 & -2 \\ -1 & 3 \end{bmatrix}$$

Step 2.

$$\text{Adj } A = \begin{bmatrix} 4 & -1 \\ -2 & 3 \end{bmatrix}$$

Step 3.

$$|A| = \begin{vmatrix} 3 & 1 \\ 2 & 4 \end{vmatrix} = 12 - 2 = 10$$

Step 4.

$$A^{-1} = \begin{bmatrix} \frac{4}{10} & \frac{-1}{10} \\ \frac{-2}{10} & \frac{3}{10} \end{bmatrix} = \begin{bmatrix} \cdot 4 & -\cdot 1 \\ -\cdot 2 & \cdot 3 \end{bmatrix}$$

This result may be checked by multiplying $A.A^{-1}$

$$= \begin{bmatrix} 3 & 1 \\ 2 & 4 \end{bmatrix} \begin{bmatrix} \cdot 4 & -\cdot 1 \\ -\cdot 2 & \cdot 3 \end{bmatrix}$$

$$= \begin{bmatrix} 3(\cdot 4)+(-\cdot 2) & 3(-\cdot 1)+(\cdot 3) \\ 2(\cdot 4)+4(-\cdot 2) & 2(-\cdot 1)+4(\cdot 3) \end{bmatrix}$$

$$= \begin{bmatrix} 1 & 0 \\ 0 & 1 \end{bmatrix}$$ The identity matrix.

(b) Find the inverse of the 3×3 matrix

$$A = \begin{bmatrix} 1 & 2 & 3 \\ 2 & 4 & 5 \\ 3 & 5 & 6 \end{bmatrix}$$

Step 1.

$$\begin{bmatrix} \begin{vmatrix} 4 & 5 \\ 5 & 6 \end{vmatrix} & -\begin{vmatrix} 2 & 5 \\ 3 & 6 \end{vmatrix} & \begin{vmatrix} 2 & 4 \\ 3 & 5 \end{vmatrix} \\ -\begin{vmatrix} 2 & 3 \\ 5 & 6 \end{vmatrix} & \begin{vmatrix} 1 & 3 \\ 3 & 6 \end{vmatrix} & -\begin{vmatrix} 1 & 2 \\ 3 & 5 \end{vmatrix} \\ \begin{vmatrix} 2 & 3 \\ 4 & 5 \end{vmatrix} & -\begin{vmatrix} 1 & 3 \\ 2 & 5 \end{vmatrix} & \begin{vmatrix} 1 & 2 \\ 2 & 4 \end{vmatrix} \end{bmatrix}$$

$$= \begin{bmatrix} -1 & 3 & -2 \\ 3 & -3 & 1 \\ -2 & 1 & 0 \end{bmatrix}$$

Step 2.

$$\text{Adj } A = \begin{bmatrix} -1 & 3 & -2 \\ 3 & -3 & 1 \\ -2 & 1 & 0 \end{bmatrix}$$

Note that in a symmetric matrix the transpose of the matrix A = A.

Step 3.

$$|A| = \begin{vmatrix} 1 & 2 & 3 \\ 2 & 4 & 5 \\ 3 & 5 & 6 \end{vmatrix}$$

$$= 1(-1) - 2(-3) + 3(-2)$$

$$= -1$$

Step 4.

$$A^{-1} = \begin{bmatrix} 1 & -3 & 2 \\ -3 & 3 & -1 \\ 2 & -1 & 0 \end{bmatrix}$$

The check by multiplication is left to the reader.

6. Equations in Matrix Notation

A set of simultaneous equations can be expressed very concisely by using matrix notation. Consider the following pair of equations:

$$\begin{aligned} 5x + 2y &= -5 \\ -3x + 4y &= 29 \end{aligned}$$

This could be written

$$\begin{bmatrix} 5 & 2 \\ -3 & 4 \end{bmatrix} \begin{bmatrix} x \\ y \end{bmatrix} = \begin{bmatrix} -5 \\ 29 \end{bmatrix}$$

(Check that multiplying the vector with the unknowns x and y by the coefficient matrix produces the left hand side of the two equations.)

Similarly

$$\begin{aligned} x + 2y + 3z &= 1 \\ 2x + 3y + 4z &= 2 \\ x - y - z &= 0. \end{aligned}$$

could be written

$$\begin{bmatrix} 1 & 2 & 3 \\ 2 & 3 & 4 \\ 1 & -1 & -1 \end{bmatrix} \begin{bmatrix} x \\ y \\ z \end{bmatrix} = \begin{bmatrix} 1 \\ 2 \\ 0 \end{bmatrix}$$

In general, if there is a set of n equations with n unknowns

$$\begin{aligned} a_{11}x_1 + a_{12}x_2 + a_{13}x_3 \dots a_{1n}x_n &= h_1 \\ a_{21}x_1 + a_{22}x_2 + a_{23}x_3 \dots a_{2n}x_n &= h_2 \\ a_{31}x_1 + a_{32}x_2 + a_{33}x_3 \dots a_{3n}x_n &= h_3 \\ \dots\dots\dots\dots\dots\dots\dots\dots \\ a_{n1}x_1 + a_{n2}x_2 + a_{n3}x_3 \dots a_{nn}x_n &= h_n \end{aligned}$$

they can be written

$$\begin{bmatrix} a_{11} & a_{12} & a_{13} & \dots & a_{1n} \\ a_{21} & a_{22} & a_{23} & \dots & a_{2n} \\ a_{31} & a_{32} & a_{33} & \dots & a_{3n} \\ \dots & \dots & \dots & \dots & \dots \\ a_{n1} & a_{n2} & a_{n3} & \dots & a_{nn} \end{bmatrix} \begin{bmatrix} x_1 \\ x_2 \\ x_3 \\ \dots \\ x_n \end{bmatrix} = \begin{bmatrix} h_1 \\ h_2 \\ h_3 \\ \dots \\ h_n \end{bmatrix}$$

or, more briefly

$$Ax = h$$

Where A is the $n \times n$ coefficient matrix
x is the $n \times 1$ vector of unknowns
h is the $n \times 1$ vector of constants.

7. The Solution of Simultaneous Linear Equations by Cramer's Rule

(i) *Simultaneous equations with two unknowns*

Let $a_{11}x + a_{12}y = h_1$ .....(1)
$a_{21}x + a_{22}y = h_2$ ....(2)

be two equations in x and y. These equations can be solved (see Part One) by multiplying (1) by a_{22} and (2) by a_{12} to eliminate y.

$$\begin{aligned} a_{11}a_{22}x + a_{12}a_{22}y &= h_1a_{22} \quad \ldots\ldots (1) \times a_{22} \\ a_{12}a_{21}x + a_{12}a_{22}y &= h_2a_{12} \quad \ldots\ldots (2) \times a_{12} \end{aligned}$$

$$(a_{11}a_{22} - a_{12}a_{21})x = (h_1a_{22} - h_2a_{12})$$

$$x = \frac{h_1a_{22} - h_2a_{12}}{a_{11}a_{22} - a_{12}a_{21}}$$

or, in determinant notation

$$x = \frac{\begin{vmatrix} h_1 & a_{12} \\ h_2 & a_{22} \end{vmatrix}}{\begin{vmatrix} a_{11} & a_{12} \\ a_{21} & a_{22} \end{vmatrix}}$$

Similarly, multiplying (1) by a_{21} and (2) by a_{11} to eliminate x, we have

$$y = \frac{\begin{vmatrix} a_{11} & h_1 \\ a_{21} & h_2 \end{vmatrix}}{\begin{vmatrix} a_{11} & a_{12} \\ a_{21} & a_{22} \end{vmatrix}}$$

Inspection of these results shows that the denominator of both x and y is the determinant of the coefficient matrix A. In the numerator of x, the coefficients of x have been replaced in the coefficient matrix A by the column of constants h_1 and h_2. In the numerator of y, the coefficients of y in A have been replaced by h_1 and h_2.

These results provide a rapid method for solving a pair of linear equations with two unknowns.

Example

Solve for x and y

$$\begin{aligned} 3x - 4y &= 5 \\ x + 7y &= 10 \end{aligned}$$

$$x = \frac{\begin{vmatrix} 5 & -4 \\ 10 & 7 \end{vmatrix}}{\begin{vmatrix} 3 & -4 \\ 1 & 7 \end{vmatrix}} = \frac{35 - (-40)}{21 - (-4)}$$

$$= \frac{75}{25}$$

$$= 3.$$

$$y = \frac{\begin{vmatrix} 3 & 5 \\ 1 & 10 \end{vmatrix}}{\begin{vmatrix} 3 & -4 \\ 1 & 7 \end{vmatrix}} = \frac{30 - 5}{21 - (-4)}$$

$$= \frac{25}{25}$$

$$= 1.$$

(ii) *Cramer's Rule*

The method for solving a pair of linear equations with two unknowns may be extended to the general case of n equations with n unknowns. If

$$Ax = h$$

(using the notation of 6) then by Cramer's Rule the solution of the equations is given by

$$x_1 = \frac{|A_1|}{|A|}$$

where $|A|$ is the determinant of the coefficient matrix A, and $|A_1|$ represents the determinant of the coefficient matrix A with the coefficients of x_1 replaced by the column of constants $h_1\ h_2\ h_3 \ldots h_n$.

$$x_2 = \frac{|A_2|}{|A|}$$

and in general

$$x_j = \frac{|A_j|}{|A|}$$

Note. If the matrix A is singular, that is $|A| = 0$ Cramer's Rule cannot be used. See below p. 191.

Example

Solve the following equations, using Cramer's Rule.

$$\begin{aligned} x + y + z &= 6 \\ 5x - y + 2z &= 9 \\ 3x + 6y - 5z &= 0 \end{aligned}$$

Writing the equations in matrix notation:

$$\begin{bmatrix} 1 & 1 & 1 \\ 5 & -1 & 2 \\ 3 & 6 & -5 \end{bmatrix} \begin{bmatrix} x \\ y \\ z \end{bmatrix} = \begin{bmatrix} 6 \\ 9 \\ 0 \end{bmatrix}$$

then,

$$x = \frac{\begin{vmatrix} 6 & 1 & 1 \\ 9 & -1 & 2 \\ 0 & 6 & -5 \end{vmatrix}}{\begin{vmatrix} 1 & 1 & 1 \\ 5 & -1 & 2 \\ 3 & 6 & -5 \end{vmatrix}} = \frac{57}{57} = 1$$

$$y = \frac{\begin{vmatrix} 1 & 6 & 1 \\ 5 & 9 & 2 \\ 3 & 0 & -5 \end{vmatrix}}{\begin{vmatrix} 1 & 1 & 1 \\ 5 & -1 & 2 \\ 3 & 6 & -5 \end{vmatrix}} = \frac{114}{57} = 2$$

$$z = \frac{\begin{vmatrix} 1 & 1 & 6 \\ 5 & -1 & 9 \\ 3 & 6 & 0 \end{vmatrix}}{\begin{vmatrix} 1 & 1 & 1 \\ 5 & -1 & 2 \\ 3 & 6 & -5 \end{vmatrix}} = \frac{171}{57} = 3$$

(iii) *An alternative formulation of Cramer's Rule*

The following set of equations

$$\begin{aligned} a_{11}x_1 + a_{12}x_2 + a_{13}x_3 &= h_1 \\ a_{21}x_1 + a_{22}x_2 + a_{23}x_3 &= h_2 \\ a_{31}x_1 + a_{32}x_2 + a_{33}x_3 &= h_3 \end{aligned}$$

yields (by Cramer's Rule)

$$x_1 = \frac{\begin{vmatrix} h_1 & a_{12} & a_{13} \\ h_2 & a_{22} & a_{23} \\ h_3 & a_{32} & a_{33} \end{vmatrix}}{|A|}$$

Expanding by the first column

$$x_1 = \frac{h_1\begin{vmatrix} a_{22} & a_{23} \\ a_{32} & a_{33} \end{vmatrix} + h_2(-1)\begin{vmatrix} a_{12} & a_{13} \\ a_{32} & a_{33} \end{vmatrix} + h_3\begin{vmatrix} a_{12} & a_{13} \\ a_{22} & a_{23} \end{vmatrix}}{|A|}$$

$$= \frac{h_1\alpha_{11} + h_2\alpha_{21} + h_3\alpha_{31}}{|A|}$$

(Where α_{11}, α_{21} and α_{31} are the *cofactors* of a_{11}, a_{21} and a_{31})

Similarly

$$x_2 = \frac{h_1\alpha_{12} + h_2\alpha_{22} + h_3\alpha_{32}}{|A|}$$

$$x_3 = \frac{h_1\alpha_{13} + h_2\alpha_{23} + h_3\alpha_{33}}{|A|}$$

Cramer's Rule may therefore be formulated in terms of the cofactors of A.

$$x_j = \frac{h_1\alpha_{1j} + h_2\alpha_{2j} \; . \; . \; . \; . \; h_n\alpha_{nj}}{|A|}$$

$$= \frac{\sum_{k=1}^{n} h_k \alpha_{kj}}{|A|} \qquad (j = 1,2 \ldots n)$$

8. The Solution of Simultaneous Linear Equations by the Inverse Method

The inverse of a matrix may be used to solve a system of simultaneous equations. In the notation of paragraph 6, a system of linear equations may be written as follows:

$$Ax = h.$$

If both sides are multiplied by the inverse of A, we have

$$A^{-1} Ax = A^{-1}h$$

or

$$x = A^{-1}h.$$

If we let the elements in the inverse A^{-1} be denoted by a^{ij}, we have

$$\begin{bmatrix} x_1 \\ x_2 \\ x_3 \\ \cdot\cdot \\ x_n \end{bmatrix} = \begin{bmatrix} a^{11} & a^{12} & a^{13} & \ldots & a^{1n} \\ a^{21} & a^{22} & a^{23} & \ldots & a^{2n} \\ a^{31} & a^{32} & a^{33} & \ldots & a^{3n} \\ \ldots & \ldots & \ldots & \ldots & \ldots \\ a^{n1} & a^{n2} & a^{n3} & \ldots & a^{nn} \end{bmatrix} \begin{bmatrix} h_1 \\ h_2 \\ h_3 \\ \cdot \\ h_n \end{bmatrix}$$

Multiplying out the right hand side of the equation gives:

$$\begin{bmatrix} x_1 \\ x_2 \\ x_3 \\ \cdot\cdot \\ x_n \end{bmatrix} = \begin{bmatrix} a^{11}h_1 + a^{12}h_2 + a^{13}h_3 \ldots a^{1n}h_n \\ a^{21}h_1 + a^{22}h_2 + a^{23}h_3 \ldots a^{2n}h_n \\ a^{31}h_1 + a^{32}h_2 + a^{33}h_3 \ldots a^{3n}h_n \\ \ldots\ldots\ldots\ldots\ldots\ldots \\ a^{n1}h_1 + a^{n2}h_2 + a^{n3}h_3 \ldots a^{nn}h_n \end{bmatrix}$$

$$= \begin{bmatrix} \sum_{k=1}^{n} a^{1k}h_k \\ \sum_{k=1}^{n} a^{2k}h_k \\ \sum_{k=1}^{n} a^{3k}h_k \\ \ldots\ldots \\ \sum_{k=1}^{n} a^{nk}h_k \end{bmatrix}$$

This general expression gives us the solution for each of the unknowns $(x_1, x_2, x_3, \ldots x_n)$. For example

$$x_5 = \sum_{k=1}^{n} a^{5k} h_k$$

Examples

(a) Solve for x_1 and x_2 in the following equations:

$$3x_1 + x_2 = 5$$
$$2x_1 + 4x_2 = 4.$$

$$\begin{bmatrix} 3 & 1 \\ 2 & 4 \end{bmatrix} \begin{bmatrix} x_1 \\ x_2 \end{bmatrix} = \begin{bmatrix} 5 \\ 4 \end{bmatrix}$$

The inverse of the coefficient matrix was worked out in example (a) paragraph 5 (iii) to be

$$\begin{bmatrix} \cdot 4 & -\cdot 1 \\ -\cdot 2 & \cdot 3 \end{bmatrix}$$

therefore

$$\begin{bmatrix} x_1 \\ x_2 \end{bmatrix} = \begin{bmatrix} \cdot 4 & -\cdot 1 \\ -\cdot 2 & \cdot 3 \end{bmatrix} \begin{bmatrix} 5 \\ 4 \end{bmatrix}$$

$$= \begin{bmatrix} 1\cdot 6 \\ \cdot 2 \end{bmatrix}$$

Answer $x_1 = 1\cdot 6$

$x_2 = \cdot 2$

(b) Solve the following system of equations.

$$\begin{bmatrix} 1 & 2 & 3 \\ 2 & 4 & 5 \\ 3 & 5 & 6 \end{bmatrix} \begin{bmatrix} x_1 \\ x_2 \\ x_3 \end{bmatrix} = \begin{bmatrix} 5 \\ 4 \\ 3 \end{bmatrix}$$

The inverse of this coefficient matrix was calculated in 5 (iii) example (b). Using this result we have

$$\begin{bmatrix} x_1 \\ x_2 \\ x_3 \end{bmatrix} = \begin{bmatrix} 1 & -3 & 2 \\ -3 & 3 & -1 \\ 2 & -1 & 0 \end{bmatrix} \begin{bmatrix} 5 \\ 4 \\ 3 \end{bmatrix}$$

$$= \begin{bmatrix} -1 \\ -6 \\ 6 \end{bmatrix}$$

Answer $x_1 = -1$
$x_2 = -6$
$x_3 = 6$

9. Singular Matrices and Systems of Simultaneous Equations

(i) *Cramer's Rule and Singular Matrices*

Using the notation of 7 (iii) for Cramer's Rule

$$x_j = \frac{h_1 \alpha_{1j} + h_2 \alpha_{2j} \; . \; . \; . \; . \; . + h_n \alpha_{nj}}{|A|} \; .$$

If, however, the matrix A is singular, then $|A| = 0$, and x_j is equal to some number divided by zero. It is impossible to divide by zero, therefore Cramer's Rule does not apply if the coefficient matrix A is singular.

(ii) *The Inverse Method and Singular Matrices*

The last step to be followed in calculating the inverse of a matrix (Step 4 in 6 (iii) above) consists of dividing every element in the adjoint matrix by $|A|$. If the matrix A is singular, this again means division by zero. Since division by zero is impossible, a singular matrix has no inverse and the inverse method of solving simultaneous equations cannot be used if the coefficient matrix A is singular.

(iii) *Two possibilities*

If the coefficient matrix A is singular, two possibilities exist:

(a) There may be *no solution* to the system of simultaneous equations. This possibility exists if, and only if,

$$h_1 \alpha_{1j} + h_2 \alpha_{2j} \; . \; . \; . \; . + h_n \alpha_{nj} \neq 0 \qquad (\text{for } j = 1, 2, \ldots \ldots n).$$

and Cramer's Rule yields

$$x_j = \frac{|A_j|}{0} \qquad (A_j \neq 0)$$

In this case, the equations are said to be inconsistent. (See IV i)

Example

Solve:

$$\begin{aligned} 3x_1 + x_2 + 2x_3 &= 2 \\ 2x_1 + 3x_2 + x_3 &= 2 \\ 5x_1 + 4x_2 + 3x_3 &= 3. \end{aligned}$$

or

$$\begin{bmatrix} 3 & 1 & 2 \\ 2 & 3 & 1 \\ 5 & 4 & 3 \end{bmatrix} \begin{bmatrix} x_1 \\ x_2 \\ x_3 \end{bmatrix} = \begin{bmatrix} 2 \\ 2 \\ 3 \end{bmatrix}$$

The determinant of the coefficient matrix (expanding by the first column) is

$$3\begin{vmatrix} 3 & 1 \\ 4 & 3 \end{vmatrix} -2\begin{vmatrix} 1 & 2 \\ 4 & 3 \end{vmatrix} +5\begin{vmatrix} 1 & 2 \\ 3 & 1 \end{vmatrix} = 0$$

The matrix is singular, Applying Cramer's Rule, we obtain

$$x_1 = \frac{2\begin{vmatrix} 3 & 1 \\ 4 & 3 \end{vmatrix} -2\begin{vmatrix} 1 & 2 \\ 4 & 3 \end{vmatrix} +3\begin{vmatrix} 1 & 2 \\ 3 & 1 \end{vmatrix}}{0} = \frac{5}{0}$$

$$x_2 = \frac{2\begin{vmatrix} 2 & 1 \\ 5 & 3 \end{vmatrix} -2\begin{vmatrix} 3 & 2 \\ 5 & 3 \end{vmatrix} +3\begin{vmatrix} 3 & 2 \\ 2 & 1 \end{vmatrix}}{0} = \frac{1}{0}$$

$$x_3 = \frac{2\begin{vmatrix} 2 & 3 \\ 5 & 4 \end{vmatrix} -2\begin{vmatrix} 3 & 1 \\ 5 & 4 \end{vmatrix} +3\begin{vmatrix} 3 & 1 \\ 2 & 3 \end{vmatrix}}{0} = \frac{-7}{0}$$

There is no solution to the system of equations. The reason for this may be seen clearly if the method of Part One is adopted.

$$\begin{aligned} 3x_1 + x_2 + 2x_3 &= 2 \quad \ldots\ldots (1) \\ 2x_1 + 3x_2 + x_3 &= 2 \quad \ldots\ldots (2) \\ 5x_1 + 4x_2 + 3x_3 &= 3 \quad \ldots\ldots (3) \end{aligned}$$

Subtracting (1) from (3) gives

$$2x_1 + 3x_2 + x_3 = 1.$$

This equation is clearly inconsistent with equation (2).

(b) There may be *many solutions* to the system of simultaneous equations. This possibility exists if, and only if,

$$h_1\alpha_{1j} + h_2\alpha_{2j} \; . \; . \; . \; . + h_n\alpha_{nj} = 0$$

$$(\text{for } j = 1, 2 \; . \; . \; . \; . \; n)$$

and Cramer's Rule yields

$$x_j = \frac{0}{0}$$

Example

Solve

$$\begin{aligned} 3x_1 + x_2 + 2x_3 &= 1 \\ 2x_1 + 3x_2 + x_3 &= 2 \\ 5x_1 + 4x_2 + 3x_3 &= 3 \end{aligned}$$

or

$$\begin{pmatrix} 3 & 1 & 2 \\ 2 & 3 & 1 \\ 5 & 4 & 3 \end{pmatrix} \begin{pmatrix} x_1 \\ x_2 \\ x_3 \end{pmatrix} = \begin{pmatrix} 1 \\ 2 \\ 3 \end{pmatrix}$$

The matrix of coefficients is the same as in the preceding example and is therefore singular. The constants on the right hand side of the equation are different, and Cramer's Rule gives:

$$x_1 = \frac{1\begin{vmatrix} 3 & 1 \\ 4 & 3 \end{vmatrix} - 2\begin{vmatrix} 1 & 2 \\ 4 & 3 \end{vmatrix} + 3\begin{vmatrix} 1 & 2 \\ 3 & 1 \end{vmatrix}}{0} = \frac{0}{0}$$

$$x_2 = \frac{1\begin{vmatrix} 2 & 1 \\ 5 & 3 \end{vmatrix} - 2\begin{vmatrix} 3 & 2 \\ 5 & 3 \end{vmatrix} + 3\begin{vmatrix} 3 & 2 \\ 2 & 1 \end{vmatrix}}{0} = \frac{0}{0}$$

$$x_3 = \frac{1\begin{vmatrix} 2 & 3 \\ 5 & 4 \end{vmatrix} - 2\begin{vmatrix} 3 & 1 \\ 5 & 4 \end{vmatrix} + 3\begin{vmatrix} 3 & 1 \\ 2 & 3 \end{vmatrix}}{0} = \frac{0}{0}$$

There are many solutions to this system of equations. This can be demonstrated if, as in the preceding example, we subtract equation (1) from equation (3). The answer is

$$2x_1 + 3x_2 + x_3 = 2$$

which is identical to equation (2). Any set of values that satisfies (1) and (3) will also satisfy (2).

$$\begin{aligned} 3x_1 + x_2 + 2x_3 &= 1 \quad \ldots\ldots (1) \\ 5x_1 + 4x_2 + 3x_3 &= 3 \quad \ldots\ldots (3) \end{aligned}$$

Solve for x_1, x_2 in terms of x_3

$$\begin{aligned} 3x_1 + x_2 &= 1 - 2x_3 \quad \ldots\ldots (1) \\ 5x_1 + 4x_2 &= 3 - 3x_3 \quad \ldots\ldots (3) \end{aligned}$$

Multiplying (1) by 4

$$\begin{aligned} 12x_1 + 4x_2 &= 4 - 8x_3 \quad \ldots\ldots (1)\times 4 \\ 5x_1 + 4x_2 &= 3 - 3x_3 \quad \ldots\ldots (3) \\ \hline 7x_1 &= 1 - 5x_3 \end{aligned}$$

$$x_1 = \frac{1 - 5x_3}{7}$$

by a similar procedure

$$x_2 = \frac{4 + x_3}{7}$$

For any value of x_3, we can now find the corresponding values for x_1 and x_2. For example, if $x_3 = 3$

$$x_1 = \frac{1-5x_3}{7} = \frac{1 - 15}{7} = -2$$

$$x_2 = \frac{4 + x_3}{7} = \frac{4 + 3}{7} = 1.$$

The values $x_1 = 2$, $x_2 = 1$, $x_3 = 3$ satisfy all three original equations. If $x_3 = 10$, we obtain another set of solutions: $x_1 = -7$, $x_2 = 2$, $x_3 = 10$. Clearly there is an infinite number of possible solutions to this system of simultaneous equations.

10. An Economic Example: a Balance of Trade Problem solved by Cramer's Rule

This problem is a development of the one considered in V, section B. There we were concerned with the income effects of an exogenous fall in the value of the exports of the "poor" primary producing countries to the "rich" industrialized countries. Now let us look at the effect such a fall will have on the balance of trade of the primary producers. We denote changes in income in the "poor" and the "rich" countries by $\triangle Y_1$ and $\triangle Y_2$, respectively; d_1 and d_2 denote the marginal propensity to spend income on domestically produced goods, and m_1 and m_2 denote the marginal propensity to import. Finally, let $\triangle T_1$ represent the balance of trade of the "poor" countries. We have the two equations for income changes

$$\triangle Y_1 = d_1\triangle Y_1 + m_2\triangle Y_2 - a \qquad (1)$$
$$\triangle Y_2 = d_2\triangle Y_2 + m_1\triangle Y_1 \qquad (2)$$

and an equation showing the change in the balance of trade

$$\triangle T_1 = m_2\triangle Y_2 - m_1\triangle Y_1 - a \qquad (3)$$

i.e. the change in the trade balance of "poor" countries is the induced change in their exports less the induced change in their imports less the initial autonomous fall in export proceeds.

It would be possible to solve these equations by a series of substitutions; a much less cumbersome method would be to write the equations in matrix form and use Cramer's Rule.

$$\begin{bmatrix} 1-d_1 & -m_2 & 0 \\ -m_1 & 1-d_2 & 0 \\ m_1 & -m_2 & 1 \end{bmatrix} \begin{bmatrix} \triangle Y_1 \\ \triangle Y_2 \\ \triangle T_1 \end{bmatrix} = \begin{bmatrix} -a \\ 0 \\ -a \end{bmatrix}$$

Since any increase in income must be spent on domestic goods or imports or *not* be spent at all (i.e. saved),

$$d + m + s = 1,$$

and we can substitute $(m + s)$ for $(1 - d)$ in the matrix. This gives

$$\begin{bmatrix} m_1 + s_1 & -m_2 & 0 \\ -m_1 & m_2 + s_2 & 0 \\ m_1 & -m_2 & 1 \end{bmatrix} \begin{bmatrix} \triangle Y_1 \\ \triangle Y_2 \\ \triangle T_1 \end{bmatrix} = \begin{bmatrix} -a \\ 0 \\ -a \end{bmatrix}$$

The determinant of the matrix of coefficients $|A|$ is

$$|A| = (m_1 + s_1)(m_2 + s_2) - m_1m_2$$
$$= m_1s_2 + m_2s_1 + s_1s_2.$$

To solve for the change in the balance of trade ΔT_1, we substitute the vector of constants for the coefficients of ΔT_1 in the determinant $|A|$.

$$\Delta T_1 = \frac{\begin{vmatrix} m_1+s_1 & -m_2 & -a \\ -m_1 & m_2+s_2 & 0 \\ m_1 & -m_2 & -a \end{vmatrix}}{|A|}$$

Simplifying the numerator

$$\Delta T_1 = \frac{-a\,(m_2s_1 + s_1s_2)}{|A|}$$

Finally, substituting the value of $|A|$, we have

$$\Delta T_1 = (a)\left\{ - \frac{m_2s_1 + s_1s_2}{m_1s_2 + m_2s_1 + s_1s_2} \right\}$$

If, as we would expect, the m's and s's are all positive fractions, the fall in the value of the "poor" countries' exports (a) will be multiplied by a negative quantity. Therefore, we can say that for any m and s values between 0 and 1, a fall in the exports of the primary producing countries will necessarily turn the balance of trade *against* these countries *by some fraction* of the original drop in value.

Needless to say, now that the problem has been set up, it would be easy to substitute the column of constants for the columns one and two in the determinant $|A|$ and obtain a general solution for ΔY_1 and ΔY_2 respectively.

EXERCISE 17

1. Evaluate

(a) $\begin{vmatrix} 2 & 9 \\ 6 & 3 \end{vmatrix}$ (b) $\begin{vmatrix} -4 & 2 \\ -1 & 18 \end{vmatrix}$ (c) $\begin{vmatrix} 0 & -13 \\ 2 & 7 \end{vmatrix}$

2. Evaluate

(a) $\begin{vmatrix} 3 & 2 & 4 \\ 6 & 0 & 1 \\ 9 & 8 & 5 \end{vmatrix}$ (b) $\begin{vmatrix} -1 & 0 & 2 \\ 0 & 3 & 6 \\ -4 & 5 & 1 \end{vmatrix}$ (c) $\begin{vmatrix} 5 & 1 & 3 \\ -2 & 1 & 4 \\ 3 & 10 & 0 \end{vmatrix}$

3. Evaluate

(a) $\begin{vmatrix} 82 & 60 & 53 & 41 \\ 4 & 2 & 6 & 3 \\ 44 & 23 & 68 & 35 \\ 4 & 3 & 2 & 2 \end{vmatrix}$ (b) $\begin{vmatrix} 12 & 3 & 2 & 8 \\ 36 & 2 & 6 & 6 \\ 50 & 1 & 8 & 3 \\ 14 & 4 & 2 & 8 \end{vmatrix}$

4. Find the value of the nine cofactors in the following matrix:

$$\begin{bmatrix} 1 & 2 & 3 \\ 2 & 3 & 2 \\ 1 & 2 & 2 \end{bmatrix}$$

5. Give the rank of the following matrices:

(a) $\begin{bmatrix} 1 & 0 & 2 \\ 0 & 2 & 1 \\ 3 & 2 & 3 \end{bmatrix}$ (b) $\begin{bmatrix} 1 & 2 & 3 \\ 23 & 3 & 4 \\ 3 & 5 & 7 \end{bmatrix}$

(c) $\begin{bmatrix} 1 & 5 & 4 & 2 \\ 0 & 3 & 1 & 2 \end{bmatrix}$ (d) $\begin{bmatrix} 2 & 3 & 1 & -1 \\ 2 & 3 & 1 & -2 \\ 4 & 6 & 2 & -3 \end{bmatrix}$

6. Find the inverse of the following matrices and check the result by multiplication:

(a) $\begin{bmatrix} 1 & 2 \\ 3 & 4 \end{bmatrix}$ (b) $\begin{bmatrix} 9 & 1 \\ 8 & 2 \end{bmatrix}$ (c) $\begin{bmatrix} 1 & 2 & 3 \\ 1 & 3 & 4 \\ 1 & 4 & 3 \end{bmatrix}$

(d) $\begin{bmatrix} 1 & 3 & 3 \\ 1 & 3 & 4 \\ 1 & 4 & 3 \end{bmatrix}$

7. Solve by Cramer's Rule:

(a) $\begin{aligned} 2x + 3y &= 11 \\ x - 4y &= 22 \end{aligned}$ (b) $\begin{aligned} x + y + z &= 6 \\ 5x - y + 2z &= 9 \\ 3x + 6y - 5z &= 0 \end{aligned}$

(c) $2x - y + z = 7$
$3x + y - 5z = 13$
$x + y + z = 5$

(d) $3x - y + 4z = 13$
$5x + y - 3z = 5$
$x - y + z = 3$

8. Use the results of 6(b) and 6(d) to solve the following systems of equations by the inverse method:

(a) $9x_1 + x_2 = 13$
$8x_1 + 2x_2 = 16$

(b) $x_1 + 3x_2 + 3x_3 = 1$
$x_1 + 3x_2 + 4x_3 = 0$
$x_1 + 4x_2 + 3x_3 = 1$

9. Find the solution, if any, to the following systems of equations:

(a) $4x + 5y - 2z = 19$
$x + y - z = 4$
$9x + 11y - 5z = 41$

(b) $x + y + 3z = 0$
$x + 2y = 6$
$x + 3y + z = 8$

(c) $8x_1 + 4x_2 + 12x_3 = 4$
$2x_1 + x_2 + 3x_3 = 1$
$x_1 - 2x_2 + 4x_3 = -7$

XVIII. DIFFERENCE EQUATIONS

1. Introduction

The economic theory first encountered by the student is usually concerned with static analysis; that is, time plays no explicit part. For example,

$$Y = C + I + G$$
$$C = f(Y)$$

(where $Y =$ national income, $C =$ consumption,
$I =$ investment, $G =$ government expenditure)

Given these basic relationships, and values for the independent variables, the equilibrium level of income can be calculated.

The realism of this simple model may be increased by introducing the element of time. National income Y in any particular period t (e.g. a year) may be described as Y_t. Y_t may have been determined by the level of income in the preceding year (Y_{t-1}). (People base their consumption patterns on last year's disposable income; investors make their decisions on the basis of last year's profits, etc.) In this case we have the functional relationship

$$Y_t = f(Y_{t-1}).$$

Such an equation is known as a *first order difference equation.* If we know the level of income at time $_{t-1}$ we can calculate the level of income at time t. It is therefore necessary to have *one initial condition* to find Y_t.

The model could be further complicated to allow for the influence on this year's income of the level of income at time t–2 as well as t–1. This would give

$$Y_t = f(Y_{t-1}) + g(Y_{t-2}).$$

This is a *second order difference equation.* To calculate Y_t it is now necessary to know *two initial conditions*—the level of income at time t–2 and at time t–1.

Example

Calculate the level of income (Y) at time t, given the following second order difference equation and two initial conditions:

$$Y_t = 5Y_{t-1} - 4Y_{t-2}$$
$$Y_{t-1} = £50m.$$
$$Y_{t-2} = £45m.$$

Solution:

$$Y_t = 5(50) - 4(45)$$
$$= 250 - 180$$
$$= £70m.$$

In general, an nth *order difference equation* will have the form:

$$Y_t = f(Y_{t-1}) + g(Y_{t-2}) \ldots + \theta(Y_{t-n})$$

and the solution will require *n initial conditions.*

We have introduced the idea of difference equations in terms of national income analysis and used the symbols Y_t, Y_{t-1}, Y_{t-2} etc. For convenience, we shall continue to use the variable Y, but it should be realized that difference equations may be used to describe investment decisions, supply functions—any state of affairs where the conditions of one period appear to be influenced by past conditions. Dynamic models are frequently based on difference equations. A similar approach, not considered here, uses differential equations, where the successive time periods t–2, t–1, t are taken to be very close together so that the difference between them tends to 0. The method of solution may be found in W.J. Baumol, *Economic Dynamics*, a very useful book which has provided the basis of the present chapter.

2. Types of Difference Equations

(i) *Linear difference equations with constant coefficients*

Difference equations may be of varying degrees of complexity. In this Outline we shall consider only linear difference equations with constant coefficients. This means there will be no equations of the form $Y_t = 3(Y_{t-1})^2$ or $Y_t = 4\log Y_{t-1}$; all the functional relationships will be linear. The general form of the nth order equation to be considered is

$$Y_t = a_1Y_{t-1} + a_2Y_{t-2} \ldots + a_nY_{t-n} + b$$

(where a_1, a_2 .... a_n and b are constants, and a_n is not equal to 0. This last condition is necessary to prevent the term a_nY_{t-n} disappearing; if it did, the equation would not be of the nth order.)

The initial conditions for this equation must be n in number, representing Y_{t-n}, $Y_{t-(n+1)} \ldots Y_{t-2}$, Y_{t-1}. These initial conditions are often presented in the form

$Y_o = K_o$ i.e. at the beginning of the first period when $t = 0$, $Y_t = Y_o = K_o$ (a constant).

$Y_1 = K_1$

$Y_2 = K_2$

. . . . . .

etc.

(ii) *Homogeneous and non-homogeneous equations*

A difference equation is said to be *homogeneous* if it is of the form:

(a) $Y_t = aY_{t-1}$
which may also be written
$Y_t - aY_{t-1} = 0$. (First-order equation).

(b) $Y_t = a_1Y_{t-1} + a_2Y_{t-2}$
which may also be written
$Y_t - a_1Y_{t-1} - a_2Y_{t-2} = 0$ (Second-order equation.)

A difference equation is *non-homogeneous* if there is a constant term in the equation. For example

(a) $Y_t = aY_{t-1} + b$
or $Y_t - aY_{t-1} = b$. (First-order equation).

(b) $Y_t = a_1Y_{t-1} + a_2Y_{t-2} + b$
or $Y_t - a_1Y_{t-1} - a_2Y_{t-2} = b$. (Second-order equation.)

3. The Problem

It is clear that given a difference equation and the correct number of initial conditions, it is possible to calculate Y_t; it is also possible by substitution to calculate Y_{t+1}, $Y_{t+2} \ldots\ldots Y_{t+n}$.

Example

Compute Y for the three periods after the initial periods given:

$$Y_t = 2Y_{t-1} - Y_{t-2}.$$

$$Y_o = 1$$

$$Y_1 = 2.$$

By substitution of the initial conditions:

$$\begin{aligned} Y_t &= 2(2) - (1) \\ &= 3. \end{aligned}$$

For the following period we have new values for Y_{t-1} and Y_{t-2}: $Y_{t-2} = 2$, $Y_{t-1} = 3$. The difference equation gives

$$\begin{aligned} Y_t &= 2(3) - (2) \\ &= 4. \end{aligned}$$

Now, substituting 4 for Y_{t-1}, and 3 for Y_{t-2}, the difference equation gives

$$\begin{aligned} Y_t &= 2(4) - (3) \\ &= 5. \end{aligned}$$

If in the above example we were interested in income 60 years from now, it would be possible to proceed in this manner through 60 calculations. However, it would be more satisfactory if we could find a formula, given a difference equation and a set of initial conditions, which could be applied to calculate Y_t in any year without having to proceed step by step from the very beginning.

Any formula or solution must

(a) satisfy the difference equation

(b) satisfy the initial conditions.

The following paragraphs present the method of solution of first-order and second-order difference equations.

4. Homogeneous First-Order Difference Equations

(i) *A numerical example*

Let us take the following difference equation and initial condition for an example:

$$Y_t = 3Y_{t-1}$$

$$Y_0 = 4.$$

The difference equation tells us that

$$Y_1 = 3Y_0 = 3 \times 4$$

$$Y_2 = 3Y_1 = 3 \times 3 \times 4$$

$$Y_3 = 3Y_2 = 3 \times 3 \times 3 \times 4$$

$$\dots\dots$$

This suggests that the solution for the general case is

$$Y_t = 3^t \times 4.$$

(ii) *A literal solution*

Given the difference equation

$$Y_t = aY_{t-1}$$

and the initial condition

$$Y_o = K_o$$

it is clear that

$$Y_1 = aY_o$$

$$Y_2 = aY_1 = a(aY_o) = a^2Y_o$$

$$Y_3 = aY_2 = a(a^2Y_1) = a^3Y_o$$

$$Y_4 = aY_3 = a(a^3Y_o) = a^4Y_o$$

$$\dots\dots$$

This suggests that the solution for the general case is

$$Y_t = a^tY_o.$$

Substituting K_o for Y_o

$$Y_t = a^tK_o.$$

This solution shows that as a and K_o are assumed constant, income Y at any period is an exponential function of t.

(iii) *Graphical interpretation of the solution*

The solution to the difference equation

$$Y_t = aY_{t-1},$$

given the initial condition

$$Y_0 = K_0$$

is $$Y_t = a^t K_0.$$

If we assume that K_0 is a positive number, the value of Y in successive periods, sometimes known as the time-path of Y, depends on the value of a. Four possibilities are shown below.

(a) $a > 1$

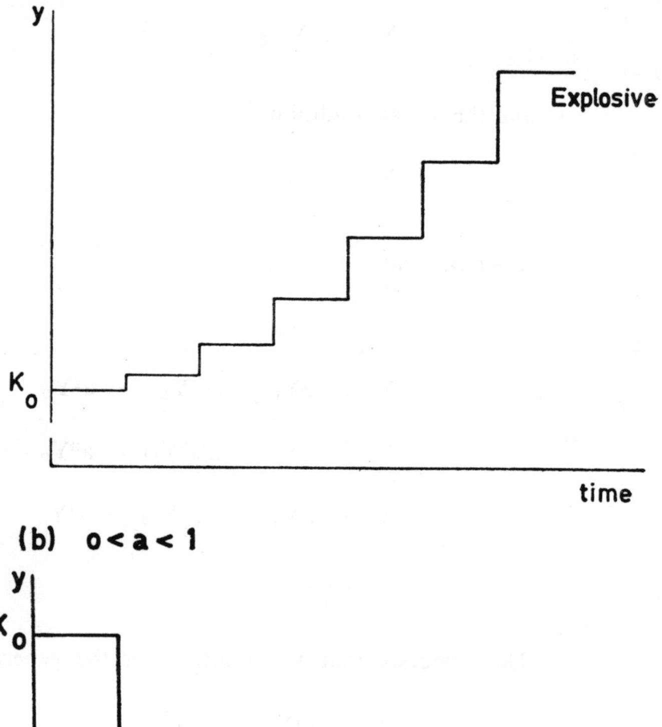

(b) $0 < a < 1$

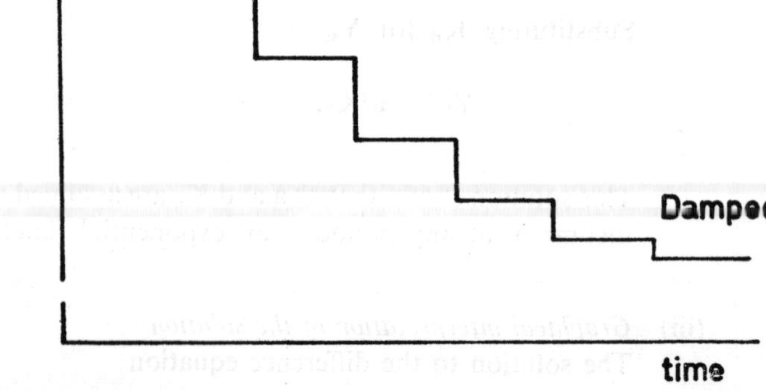

(c) $a < -1$

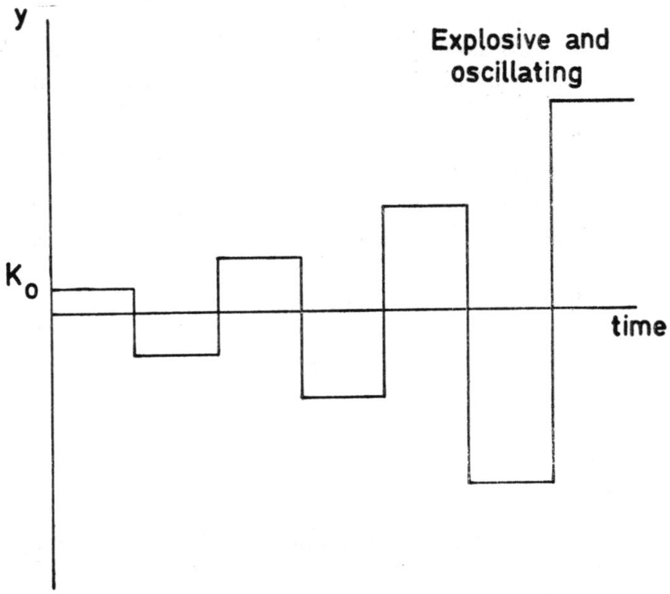

(d) $-1 < a < 0$

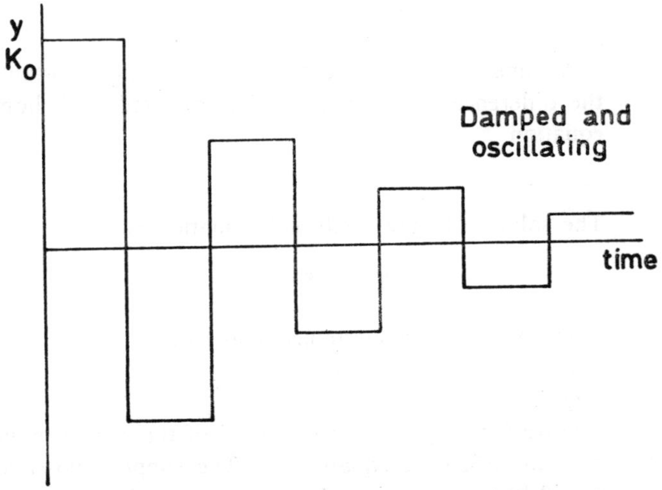

The borderline cases where $a = -1$, 0 or 1 are left to the reader. (See also question 3 in Exercise 18.)

5. The Non-homogeneous First-order Difference Equation

The solution of the first-order homogeneous equation provides a basis for the solution of more complex cases. In each case we assume that the solution will show that Y is an exponential function of t,

$$Y = A^t B$$

where A and B are undetermined constants.

The method of solving the more complex cases is to start from this assumption and then to look for values of A and B which will satisfy both the given difference equation and the initial conditions.

The procedure for solving non-homogeneous equations follows a pattern; the following illustration is therefore set out in steps. Each step is explained, but no proofs are given. The reader wishing a fuller exposition should consult Baumol (*op. cit.*) Chapters 9 and 10.

Suppose the following difference equation has been established

$$Y_t = 5Y_{t-1} - 8.$$

The initial condition is

$$Y_o = 5,$$

and we wish to find a solution for Y_t. The method is to find a "general solution"—a formula which satisfies the difference equation — and then to modify the general solution in order to take account of the initial condition.

Step 1

Assume for the moment that there is no constant term in the difference equation and solve the "reduced" homogeneous equation

$$Y_t = 5Y_{t-1}.$$

The solution to the "reduced" equation is

$$Y_t = 5^t x$$

where x is an undetermined constant.

Step 2

Solve for a "particular solution" of the complete non-homogeneous difference equation. The simplest possible solution would be the case where Y_t is always equal to a constant, which we may call z. In this particular case we should know

$$Y_t = Y_{t-1} = z.$$

Substituting z for Y_t and Y_{t-1} in the difference equation gives

$$z = 5z - 8$$
$$-4z = -8$$
$$z = 2.$$

Step 3

Combine the results of Step 1 and Step 2 to find what is known as the "general solution."

$$Y_t = 5^t x + 2.$$

Step 4

The general solution must now be modified to take account of the initial conditions. We know that $Y_o = 5$; if the general solution is to be satisfied

$$Y_o = 5^o x + 2$$

therefore

$$5^o x + 2 = 5$$

5° is, of course, equal to 1, so we have

$$\begin{aligned} x + 2 &= 5 \\ x &= 3. \end{aligned}$$

The final solution is

$$Y_t = 5^t(3) + 2.$$

To check this solution we must ascertain that it satisfies both the initial condition and the difference equation.

The initial condition that $Y_o = 5$ is satisfied: substitution in

$$Y_t = 5^t(3) + 2$$

gives

$$\begin{aligned} Y_o &= 5^o(3) + 2 \\ &= 5. \end{aligned}$$

To see whether the difference equation is satisfied, use the solution to solve for $t = 1$, $t = 2$.

$$\begin{aligned} Y_1 &= 5^1(3) + 2 = 17 \\ Y_2 &= 5^2(3) + 2 = 77. \end{aligned}$$

If the difference equation is satisfied

$$Y_t = 5Y_{t-1} - 8$$

therefore

$$\begin{aligned} Y_2 &= 5Y_1 - 8 \\ 77 &= 5(17) - 8 \\ &= 77. \end{aligned}$$

The solution is correct.

Exercise

Solve the following equation, given

$$Y_t = 10Y_{t-1} - 18$$
$$Y_0 = 8$$

Answer $$Y_t = 10^t(6) + 2$$

6. Homogeneous Second-order Difference Equation

To solve a second-order equation we again assume that the the solution will be an exponential function of t. Starting from the simplest assumption

$$Y_t = x^t$$

we then proceed to adapt it to satisfy first the difference equation, and secondly the initial conditions. For an example, let us take the difference equation

$$Y_t = 6Y_{t-1} - 8Y_{t-2}$$

with the initial conditions

$$Y_0 = 8$$
$$Y_1 = 22.$$

Step 1

Assume that the solution will be of the form

$$Y_t = x^t$$

and substitute x^t for Y_t in the difference equation:

$$x^t = 6x^{t-1} - 8x^{t-2}$$

If this equation is true for all values of t, it will be true when $t = 2$. Substitute 2 for t in the above equation to yield the quadratic equation

$$x^2 = 6x - 8.$$

Step 2

Solve this quadratic equation to find the two values of x which satisfy

$$x^2 - 6x + 8 = 0.$$

By factoring

$$(x-2)\ (x-4) = 0.$$

Either $x_1 = 2$ *or* $x_2 = 4$.

These results give us a solution for the difference equation:

$$Y_t = a(x_1)^t + b(x_2)^t$$
$$= a(2)^t + b(4)^t$$

Suitable values for the coefficients a and b must now be found.

Step 3

The values of a and b in the expression

$$Y_t = a(2)^t + b(4)^t$$

can be found by taking the two initial conditions into account. We know that

$$Y_o = 8$$
$$Y_1 = 22$$

Substituting in the above equation

$$Y_1 = a(2)^o + b(4)^o = a + b$$
$$Y_o = a(2)^1 + b(4)^1 = 2a + 4b$$

Hence

$$a + b = 8 \quad \ldots\ldots \quad (1)$$
$$2a + 4b = 22 \quad \ldots\ldots \quad (2)$$

$$2a + 2b = 16 \quad \ldots\ldots \quad (1) \times 2$$
$$2a + 4b = 22 \quad \ldots\ldots \quad (2)$$

$$-\ 2b = -6$$
$$b = 3$$

By substitution if $b = 3$, $a = 5$.

The final solution is

$$Y_t = 5(2)^t + 3(4)^t.$$

The student should check that this solution satisfies both the initial conditions and the difference equation.

Exercise

Solve the following second-order difference equation, given

$$Y_t = 5Y_{t-1} - 6Y_{t-2}$$
$$Y_o = -3$$
$$Y_1 = -4$$

Answer $Y_t = 2(3)^t - 5(2)^t$

7. Non-homogeneous Second-order Difference Equation

The method of solving a non-homogeneous second-order equation is similar to that of solving a non-homogeneous first-order equation. We first ignore the constant term and solve the "reduced" homogeneous equation. Then a "particular" solution is found and combined with the solution of the reduced equation to give a "general" solution to the difference equation. Finally the initial conditions are introduced to give the complete solution.

For example let us take the difference equation

$$Y_t = 3Y_{t-1} + 10Y_{t-2} - 12,$$

given the initial conditions

$$Y_o = 9$$
$$Y_1 = -1.$$

Step 1

Solve the "reduced" homogeneous equation

$$Y_t = 3Y_{t-1} + 10Y_{t-2}$$

by the method given in the preceding paragraph.

Assume $Y_t = x^t$, then

$$x^t = 3x^{t-1} + 10x^{t-2}$$

and $x^2 = 3 + 10.$

Solving for x_1 and x_2

$$x^2 - 3x - 10 = 0$$
$$(x-5)(2+x) = 0$$

Either $x_1 = 5$ *or* $x_2 = -2.$

We have, therefore

$$Y_t = a(5)^t + b(-2)^t$$

Step 2

Find a "particular" solution for the complete non-homogeneous difference equation. Again, take the simplest case where Y_t is always equal to a constant z. In this case

$$Y_t = Y_{t-1} = Y_{t-2} = z$$

and by substituting z for all values of Y, the difference equation becomes

$$z = 3z + 10z - 12$$
$$-12z = -12$$
$$z = 1$$

Step 3

Combine the results of Step 1 and Step 2 to find the "general solution".

$$Y_t = a(5)^t + b(-2)^t + 1$$

Step 4

Introduce the initial conditions in order to find the correct values of a and b. The initial conditions are

$$Y_o = 9$$
$$Y_1 = -1.$$

Using the "general solution"

$$Y_o = a(5)^o + b(-2)^o + 1 = a + b + 1$$

$$Y_1 = a(5)^1 + b(-2)^1 + 1 = 5a - 2b + 1.$$

Therefore

$$a + b + 1 = 9 \quad \ldots. \quad (1)$$
$$5a - 2b + 1 = -1 \quad \ldots. \quad (2)$$

$$a + b = 8 \quad \ldots. \quad (1)$$
$$5a - 2b = -2 \quad \ldots. \quad (2)$$

$$-2a - 2b = -16 \quad \ldots. \quad (1) \times (-2)$$
$$5a - 2b = -2 \quad \ldots. \quad (2)$$

$$-7a = -14$$
$$a = 2.$$

By substitution $\quad b = 6.$

The final solution is

$$Y_t = 2(5)^t + 6(-2)^t + 1.$$

Exercise

Solve the following second-order difference equation, given

$$Y_t = 6Y_{t-1} - 8Y_{t-2} - 9$$
$$Y_o = 5$$
$$Y_1 = 19$$

Answer $\quad Y_t = 3(4)^t + 5(2)^t - 3$

8. Limitation of the above Methods of Solution

All the examples chosen to illustrate the solution of first and second-order difference equations have come out very neatly.

It should be made clear that these methods will not always work.

One problem which may arise in the solution of non-homogeneous difference equations is the possibility that the "particular" solution will yield

$$z = 0,$$

which by definition of the non-homogeneous equation is not possible. For example

$$Y_t = Y_{t-1} + 5.$$

If we let $Y_t = Y_{t-1} = z$, we have

$$z = z + 5$$
$$5 = 0$$

Therefore the methods given above will not work.

Another problem lies in the solution of the second order difference equations when the roots of the derived quadratic equation are either equal or imaginary. For example

$$Y_t = 8Y_{t-1} - 16Y_{t-2}$$

yields, on the hypothesis that $Y_t = x^t$,

$$x^t = 8x^{t-1} - 16x^{t-2}$$

and $$x^2 = 8x - 16$$

The solution of

$$x^2 - 8x + 16 = 0$$

is either $x_1 = 4$ or $x_2 = 4$.

A quadratic which has no real solution comes from the difference equation

$$Y_t = 2Y_{t-1} - 5Y_{t-2}$$

The only solution to

$$x^2 - 2x + 5 = 0$$

involve the imaginary number $\sqrt{-1}$, commonly known as i.

The methods of dealing with these problems are not given here; the interested reader should consult Baumol, Chapter 10.

9. A. Cobweb Model (Using Difference Equations)

The supply and demand models discussed in Chapter V were static, that is, time played no explicit part. This simplification is often particularly unrealistic on the supply side. For any commodity which takes some time to produce, the decisions about the quantity to be supplied at any time t must have been based on knowledge of prices at time t–1 and mere guesses about prices at time t. Agricultural production provides one of the best examples of supply function of the form

$$Q_t = f(P_{t-1}).$$

Once the element of time is introduced, there is no assurance that any equilibrium price and quantity will exist in the market. Two possibilities are illustrated below.

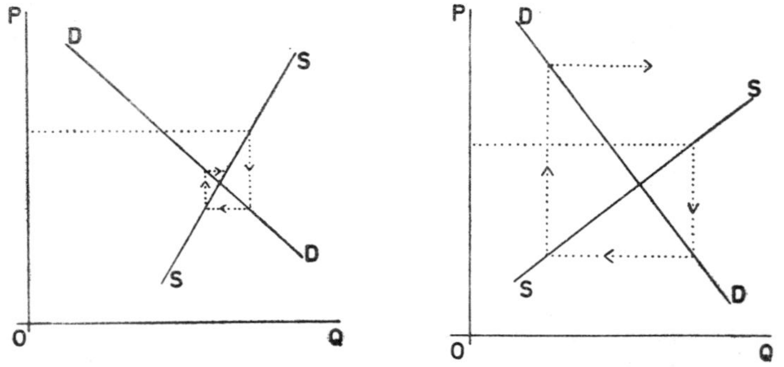

In the first case there is a tendency to equilibrium; in the second, the fluctuations in price become more and more acute.

Let us take an algebraic example. Suppose that we have the supply equation

$$Q_t^s = 3P_{t-1}$$

and the demand equation

$$Q_t^d = 40 - 2P_t.$$

We know that in the initial period (time t_0) the market price was 10. Is this a stable situation? In other words, is there some price at which the quantity demanded will be exactly equal to the quantity supplied for all periods of time? Let $Q_t^s = Q_t^d$

This gives us a difference equation in P:

$$3P_{t-1} = 40 - 2P_t.$$

Rearranging

$$P_t = \frac{-3}{2} P_{t-1} + 20$$

This is a non-homogeneous first order difference equation Solving by means of the steps:

Step 1
The solution of the "reduced" equation is

$$P_t = \left(-\frac{3}{2}\right)^t x$$

Step 2
The "particular" solution is given by

$$z = -\frac{3}{2} z + 20$$

$$\frac{5}{2} z = 20$$

$$z = 8$$

Step 3
Combining these results

$$P_t = \left(-\frac{3}{2}\right)^t x + 8$$

Step 4
Substituting the initial price $P_o = 10$

$$P_o = \left(-\frac{3}{2}\right)^o x + 8$$

$$10 = x + 8$$

$$x = 2.$$

Therefore the complete solution is

$$P_t = \left(-\frac{3}{2}\right)^t (2) + 8.$$

In this situation, prices will be very unstable (see diagram p.205) and, in mathematics at least, an equilibrium price does not exist for this particular commodity once we take time into consideration.

EXERCISE 18

1. Solve the following systems:

(a) $Y_t = 5Y_{t-1}$
$Y_o = 8$

(b) $Y_t = -3Y_{t-1}$
$Y_o = 2$

(c) $Y_t = \frac{1}{2}Y_{t-1}$
$Y_o = 4$

(d) $Y_t = -\frac{1}{4}Y_{t-1}$
$Y_o = 12.$

2. Describe the time path of Y in the equations of question (1).

3. The general form of the solution of the first-order homogeneous equation is

$$Y_t = A^tB.$$

Paragraph 4(ii) showed how different values of A affected the time path of Y if B was assumed positive. Give B a negative value and analyse the effect on the time path of Y for different values of A.

4. Solve the following systems:

(a) $Y_t = 4Y_{t-1} - 9$
$Y_o = 5$

(b) $Y_t = 6Y_{t-1} - 10$
$Y_o = 13$

(c) $Y_t = -3Y_{t-1} + 20$
$Y_o = -8$

5. (a) By following the steps for the solution of a first-order non-homogeneous equation, show that given
$Y_t = aY_{t-1} + b$
$Y_o = K_o$
the final solution is

$$Y_t = a^t(K_o - \frac{b}{1-a}) + \frac{b}{1-a}$$

(b) Apply this formula to one of the problems in question (4).

6. Solve the following systems.

(a) $Y_t = 4Y_{t-1} - 3Y_{t-2}$
$Y_0 = 2$
$Y_1 = 4.$

(b) $Y_t = 3Y_{t-1} + 10Y_{t-2}$
$Y_0 = 8$
$Y_1 = -2$

(c) $Y_t = 8Y_{t-1} - 12Y_{t-2}$
$Y_0 = 57$
$Y_1 = 334$

(d) $Y_t = 4Y_{t-2}$
$Y_0 = 8$
$Y_1 = -4$

7. Solve the following systems:

(a) $Y_t = 2Y_{t-1} + 3Y_{t-2} + 8$
$Y_0 = 6$
$Y_1 = 2$

(b) $Y_t = 3Y_{t-1} + 18Y_{t-2} - 40$
$Y_0 = 7$
$Y_1 = 14$

(c) $Y_t = 5\frac{1}{4}Y_{t-1} - 5Y_{t-2} + 3$
$Y_0 = 9$
$Y_1 = 18\frac{1}{2}.$

(d) $Y_t = 4Y_{t-2} + 90$
$Y_0 = -32$
$Y_1 = -14.$

8. Explain why the given methods do not work for

(a) $Y_t = Y_{t-1} - 3$

(b) $Y_t = 4Y_{t-1} - 3Y_{t-2} + 8$

(c) $Y_t = 6Y_{t-1} - 9Y_{t-2}$

(d) $Y_t = 3Y_{t-1} - 10Y_{t-2} + 4.$

ANSWERS TO EXERCISES

Exercise 1.

A.
1. $-2a-2b+4c$
2. $5x-5y-2z$
3. $-3m-12n$
4. $abc-7a-3b$

B.
1. $6x-5y+4z$
2. $13x^2+4x-7$
3. $3a+8b-2c$
4. $13x-15$
5. $12ab-5bc+4cd$

C.
1. $4x-4y+z$
2. $a+3b+9c$
3. $-5a+6b+c$
4. $3x^3+7x^2-2x-16$
5. $-ab+ac-5$

D.
1. x^4y
2. $-x^3y^3$
3. $-a^4-a^3+a^2$
4. $a^3-3a^2b+3ab^2-b^3$
5. $6x^2+xy-y^2$

E.
1. m^2-2
2. $-c^2+c-1$
3. $-2x^2+4x+10$
4. $2x-4y^2$
5. $a-12$

F.
1. 1
2. 5
3. 13
4. -5
5. -17
6. -3

Exercise 2.

A.
1. $(x-5)(x+2)$
2. $(p-3)^2$
3. $3xy(2x^2-5x-1)$
4. Prime
5. $(p-8)(p+7)$
6. $(r-12)(r+5)$
7. $(3-2x)(3+2x)$
8. $(a-7)^2$
9. $(c-5)(2c-7)$
10. $(a-8b)(a+8b)$
11. $(x+6)(x+2)$
12. $a(12a-35b)$
13. $(a-3b^2)^2$
14. $(4x+y)^2$
15. Prime

B. 1. $\dfrac{x^3-y^3}{x^2y^2}$

2. $\dfrac{4x}{(1-x)\ (1+x)}$

3. $\dfrac{5x}{(2x-1)\ (x+4)}$

4. $\dfrac{a+6}{(a-2)\ (a+2)}$

C. 1. $\frac{1}{2}$

2. $\dfrac{3\ (x-1)}{4x}$

3. $4\ (x^2+y^2)$

4. 1

D. 1. mn

2. $\dfrac{a^2-2a+3}{a+1}$

Exercise 3

A. 1. $x=-2$

2. $x=4$

3. $x=3$

4. $y=4$

B. 1. $c=12$

2. $x=-1$

3. $x=-2$

4. $x=0$

C. 1. $c=13$ or -1

2. $x=5$ or -5

3. $x=3$ or -4

4. $x=9$ or -13

D. 1. $x=\frac{1}{5}$ or -2

2. $b=24$ or -4

3. $x=5$ or 2

4. $x=4.414$ or 1.586

E. 1. $y=3a$
2. $y=\frac{a+b^2}{3b}$
3. $y=\frac{-b\pm\sqrt{b^2-4ac}}{2a}$
4. $y=\sqrt{2-2b-b^2}$
5. $y=a$ or $2a$

F. 1. $a=4$; $b=-6$
2. $x=\frac{5}{6}$; $y=-4$
3. $x=3000$; $y=5000$
4. $x=-1$; $y=-6$; $z=6$
5. $x=0$; $y=1$
$x=3$; $y=4$
6. $x=-19$; $y=\frac{5}{3}$
$x=-5$; $y=11$

G. 1. (*a*) $3\leq x\leq 7$
(*b*) $0<y\leq 10$
(*c*) $x-y>2$
2. (*a*) $x\geq 4$
(*b*) $-1<x<11$
(*c*) $-16\leq x\leq 4$
3. Proof.

Exercise 4.

1. (*a*) $x=\frac{4}{3}y$; $y=\frac{3}{4}x$
(*b*) $x=5y-10$; $y=\frac{x+2}{5}$
(*c*) $x=\sqrt{2y-6}$; $y=\frac{x^2+3}{2}$
(*d*) $x=\sqrt[3]{4y^2}$; $y=\sqrt{\frac{x^3}{2}}$

2. (*a*) parabolic; linear; hyperbolic; cubic; parabolic
(*b*) 0; -1; ∞; 4; 15
(*c*) large and positive; large and positive; small and positive; large and positive; large and negative.
(*d*) large and positive; large and negative; small and negative; large and negative; large and negative.
3. (*a*) Graphs are parallel.
(*b*) There are three solutions.
4. $(-2, 2)$ and $(4, 14)$
5. $(2, 12)$ and $(8, 3)$
6. $x=-2.45$; 1; 2.45

Exercise 5

SECTION A

1. Demand; supply; supply; demand; supply; demand.
   Assumptions: As price rises, quantity demanded drops.
   As price rises, quantity supplied rises.

2. (*a*) P=3; Q=40
   (*b*) P=20; Q=600
3. (*a*) P=8.54; Q=36.4
   (*b*) P=12.1 Q=73.6
4. (*a*) $Q^d = 120 - 12P$
   $Q^s = 9P - 27$
   (*b*) P=7; Q=36
5. (*a*) If perfect competition:—
   $MC_1 = MC_2 = \text{Price}$
   (*b*) (i) 3480
   (ii) 1020
   (*c*) Shs. 21.40
6. (*a*) 50x+25y=2000
   (*b*) Slope shows rate of substitution between x and y, given relative prices.

SECTION B.

1. (*a*) Y=C+I+G
   Y=1200; C=1095; I=80; G=25
   (*b*) 1200; 700

2. (*a*) For equilibrium: I=S
   Y=1400; S=250
   (*c*) (i) Y=1100
   (ii) Y=1600
3. C=110+.85Y
4. (*a*) S=.2Y−30
5. (*b*) G=50
   (*c*) 4
6. (*c*) (i) 315 (ii) 5% (iii) 2000; 2100 (iv) 5%.
7. $$\triangle Y_1 = \frac{-a\,(1-d_2)}{(1-d_1)\,(1-d_2)-m_1\,m_2}$$
   $$\triangle Y_2 = \frac{-am_1}{(1-d_1)\,(1-d_2)-m_1m_2}$$

Exercise 6

A. 1. 3
2. $4x$
3. $3x^2$
4. $2x-2$

B. 1. $7x^6$
2. $6x+2$
3. $10x-3$
4. $18x^2+x-1$
5. $\frac{-4}{x^5}$
6. $\frac{-12}{x^5}$
7. $\frac{1}{2\sqrt{x}}$
8. $\frac{-1}{2\sqrt{x^3}}$
9. $\frac{2}{\sqrt[3]{x^2}}$
10. $9x^2+4x+3$
11. $-9x^2+2x-9$
12. $10x^4-12x^2+20x$
13. $\frac{4x^2-2x-8}{(4x-1)^2}$
14. $\frac{-3x^2+8x+9}{(x^2+3)^2}$
15. $\frac{-5x^2+60x-2}{(2-5x^2)^2}$
16. $6(2x-7)^2$
17. $-14x(5-x^2)^6$
18. $5(x^2-3x)^4(2x-3)$
19. a
20. $2ax+b$
21. $3ax^2+2bx+c$

C. 1. .5
2. .2

Exercise 7

1. Minimum at $(5, -13)$
2. Minimum at $(3, -3)$

3. Maximum at (2, 12)
4. Maximum at (−3, 57)
   Minimum at (3, −51)
   Point of inflexion at (0, 3)
5. Point of inflexion at (2, 1)
6. Maximum at (0, 0)
   Minimum at (2, −4)
   Point of inflexion at (1, −2)
7. Maximum at (−1, −2)
   Minimum at (1, 2)
8. Point of inflexion at (3, −4)

Exercise 8.

1. (*a*) $\frac{\delta y}{\delta x}=2x-4z$

   $\frac{\delta y}{\delta z}=-4x+2z$

   (*b*) $\frac{\delta y}{\delta x}=2xz+4z-3z^2$

   $\frac{\delta y}{\delta z}=x^2+4x-6xz$

   (*c*) $\frac{\delta y}{\delta x}=6z+4$

   $\frac{\delta y}{\delta z}=6x+\frac{1}{z^2}$

   (*d*) $\frac{\delta y}{\delta x}=\frac{1}{\sqrt{x}}-z$

   $\frac{\delta y}{\delta z}=-x-5$

   (*e*) $\frac{\delta y}{\delta x}=14x\,(x^2-z^3)^6$

   $\frac{\delta y}{\delta z}=-21z^2\,(x^2-z^3)^6$

2. (*a*) $\triangle y \approx (4x+5z)\,(\triangle x)+(5x+6z)\,(\triangle z)$
   (*b*) $\triangle y \approx 9.1$
   (*c*) (i) $\triangle y=9.2$
   (ii) Error=.1; 1.09%
3. (*a*) $\triangle y \approx (6x-4z)\,(\triangle x)+(-4x-2z)\,(\triangle z)$
   (*b*) Proof.
4. (*a*) $\triangle y \approx (6w+2x^2)\,(\triangle w)+(4wx-5z^2)\,(\triangle x)$
   $+(-10xz+1)\,(\triangle z)$

   (*b*) $\triangle y \approx -14.6$

Exercise 9

A. 1. $\frac{x^4}{4}+C$

2. x^3+C

3. $x^4+\frac{2}{3}x^3-x+C$

4. $3x^3-2x^2-6x+C$

5. $\frac{3}{2}x^4+5x^3-6x^2+C$

6. $\frac{x^3}{3}+2x-\frac{1}{x}+C$

B. 1. 121.5

2. 60

3. 4.5

4. 1.5

C. 1. 50

2. $1\frac{1}{3}$

3. $10\frac{1}{8}$

4. $\frac{1}{128}=.00781$

D. 1. $6\frac{1}{5}$

2. $18\frac{2}{3}$

3. $110\frac{5}{6}$

4. $32\frac{1}{2}$

E. 1. (*a*) $6\,(2x+5)^2$

(*b*) $(2x+5)^3+C$

(*c*) $\frac{(ax+b)^{n+1}}{a\,(n+1)}+C$

2. $y=x^2-x+1$

3. $y=x-\frac{2}{3}x^3+1$

Exercise 10.

1. (*a*) Suitable

(*b*) Not suitable

(*c*) Suitable

(*d*) Not suitable

(*e*) Suitable

x=factor input; y=output

2. (*a*) O=50+4F
   (*b*) O=50+48F−F²
   or $O=50+24F-\frac{F^2}{2}$

3. (*a*) (i) $AP=\frac{10}{L}+12-L$
   (ii) MP=12−2L
   (*b*) (i) 6
   (ii) 46
   (iii) 0
   (iv) $7\frac{2}{3}$

4. (*a*) 0.8
   (*b*) 0.04

5. (*a*) $AP=40+3F-\frac{F^2}{3}$
   MP=40+6F−F²
   (*b*) F=10
   0=366.67
   (*c*) F=3; MP=49
   (*d*) *Either* after MP=AP, $F=4\frac{1}{2}$
   or after MP at a maximum, F=3.

6. (*a*) $\frac{\delta 0}{\delta a}=24b-20a$
   $\frac{\delta 0}{\delta b}=24a-16b$
   (*b*) 10 : 11
   (*c*) $\triangle 0 \approx (24b-20a)(\triangle a)+(24a-16b)(\triangle b)$
   (*d*) $\triangle 0 \approx 6.4$

7. (*a*) F+2M=80
   (*e*) M=10; F=20
   M=15; F=30
   M=20; F=40
   (*f*) M=20; F=40; 0=800

8. (*b*) x=22.4

9. (*a*) (i) $R^1-C^1=0$
   $R^{11}-C^{11}<0$
   (ii) Marginal revenue equals marginal cost; marginal cost increasing more rapidly than marginal revenue.
   (*b*) Q=13

10. (*a*) $x=16\frac{2}{3}$
    (*b*) The same

11. (*a*) Monopolistic
    (*b*) x=10.5; R=110.25
    (*c*) x=3
    (*d*) x=6
12. (*a*) $-1; -\frac{1}{4}; \frac{1}{2}$
    (*b*) $-3; -\frac{2}{3}$
13. (*a*) Elastic, perfectly elastic; inelastic; constant or unitary elasticity; completely inelastic.
    (*b*) (i) −1
    (ii) −1
    (iii) −1
    (*c*) (i) $\frac{4}{109}=.0367$
    (ii) $\frac{3}{58}=.0517$
    (iii) 0
14. (*a*) $\frac{dP}{P}=\frac{.15}{(e+\eta)}$
    (*b*) $\frac{1}{30}$

Exercise 11

A. 1. 37900
2. .0521
3. .0513
4. .398
5. 2.95

B. 1. Proof.
2. .6990; 2.0970
3. 3.2
4. (*a*) .356
   (*b*) .0305
5. (*a*) 2 or −4
   (*b*) −.602
6. (*a*) 2
   (*b*) 2.74

Exercise 12

1. (*a*) 2960
   (*b*) 3187
2. 2004
   (*a*) 591.7
   (*b*) 594.4

4. 753.4
5. (*a*) 3125
   (*b*) 2784
6. (*a*) 1053 or 1008
   (*b*) 1163 or 1138
7. (*a*) 77000
   (*b*) 57.6
10. $x=2.4$; $x=.4$
11. (*a*) e^{x+1}
    (*b*) $(2x-3)(e^{x^2-3x})$
    (*c*) $\frac{-1}{e^x}$
    (*d*) $2(3x^2-3e^x)(6x-3e^x)$
12. (*a*) $\frac{2}{2x+1}$
    (*b*) $\frac{-2x}{7-x^2}$
    (*c*) $\frac{4}{x}$
    (*d*) 0
13. (*a*) $e^{3x-1}+C$
    (*b*) $e^{2x}-2x^2+C$
    (*c*) $\log_e(7x-3)+C$
    (*d*) 2.3026

Exercise 13

1. (*a*) $-1, 3, 7, \ldots . 27$
   (*b*) $1, 3, 9, \ldots . 2187$
   (*c*) $-1, 1, -1, \ldots . 1$
2. -220
3. 490
4. 2, 6, 10, 14, 18, 22, 26, 30.
5. 4, 9, 14 or 14, 9, 4
6. -170
7. 3, 9, 27 or 27, 9, 3
8. 4 is the limit
9. 5000
10. Limit is $(1+a)$ provided $a > 0$ or $a < -2$

Exercise 14

1. (*a*) 1,532,000
   (*b*) 31.15%
   (*c*) 2.5%
2. (*a*) 4.6%
   (*b*) 6.8%
3. 2.9%
4. (*a*) 35
   (*b*) 28.1
   (*c*) 23.5
5. (*a*) 1.74%
6. 1971
7. Tea: 100; 115; 135; 135
   Nuts: 100; 106; 117, 128
8. Largest absolute increase 1961-2
   Largest proportionate increase 1957-8
9. (*a*) Coffee: 91.7
   Cloth: 250
   Tractors; 150
   (*b*) 125
10. (*a*) 107.2
    (*b*) 108

Exercise 15.

1. (*a*) $\frac{2}{9}$
   (*b*) $\frac{2}{3}$
   (n) $\frac{1}{12}$
2. 336; 360; 15; 6
3. (*a*) 120
   (*b*) 360
   (*c*) 3360
4. (*a*) 4536
   (*b*) 9000
5. 210
6. 154, 440
7. 35, 210; 6; 1
8. 1365
9. 210
10. 126; 70

11. $70x^4y^4$
12. 78
13. (*a*) $1+5x+10x^2+10x^3+5x^4+x^5$
    (*b*) $1-5x+10x^2-10x^3+5x^4-x^5$
14. $x^6+12x^5y+60x^4y^2+160x^3y^3+240x^2y^4+192xy^5+64y^6$; 1.12616.
15. .7931

Exercise 16.

1. (*a*) Multiplication
   (*b*) Multiplication
   (*c*) Addition
   (*d*) Multiplication
   (*e*) Multiplication.

2. (*a*) $\begin{bmatrix} 3 & 10 \\ 2 & 2 \end{bmatrix}$

   (*b*) $\begin{bmatrix} 7 & 7 & -14 \\ 9 & -3 & 3 \\ 22 & 3 & -9 \end{bmatrix}$

3. (*a*) $\begin{bmatrix} 1 & 9 & -9 \\ 8 & 3 & 8 \end{bmatrix}$

   (*b*) $\begin{bmatrix} 3 & 1 \\ 11 & -13 \\ -9 & -7 \end{bmatrix}$

4. Proofs.
5. (*a*) $\begin{bmatrix} 17 \end{bmatrix}$

(*b*)

$$\begin{bmatrix} 8 & 10 & 12 \\ 12 & 15 & 18 \\ -4 & -5 & -6 \end{bmatrix}$$

(*c*)

$$\begin{bmatrix} 3 & 8 \\ 8 & -12 \end{bmatrix}$$

(*d*)

$$\begin{bmatrix} 19 & 4 & -4 & -4 \end{bmatrix}$$

Exercise 17

1. (*a*) −48
   (*b*) −70
   (*c*) 26
2. (*a*) 126
   (*b*) 51
   (*c*) −257
3. (*a*) −108
   (*b*) −52
4. Matrix of cofactors:

$$\begin{bmatrix} 2 & -2 & 1 \\ 2 & -2 & 0 \\ -5 & 4 & -1 \end{bmatrix}$$

5. (*a*) 3
   (*b*) 3
   (*c*) 2
   (*d*) 2
6. (*a*)

$$\begin{bmatrix} -2 & 1 \\ 1.5 & -.5 \end{bmatrix}$$

(*b*)
$$\begin{bmatrix} .2 & -.1 \\ -.8 & .9 \end{bmatrix}$$

(*c*)
$$\begin{bmatrix} 3.5 & -3 & .5 \\ -.5 & 0 & .5 \\ -.5 & 1 & -.5 \end{bmatrix}$$

(*d*)
$$\begin{bmatrix} 7 & -3 & -3 \\ -1 & 0 & 1 \\ -1 & 1 & 0 \end{bmatrix}$$

7. (*a*) $x=10$; $y=-3$
   (*b*) $x=1$; $y=2$; $z=3$
   (*c*) $x=4$; $y=1$; $z=0$
   (*d*) $x=2$; $y=1$; $z=2$
8. (*a*) $x = 2$; $x_2=4$
   (*b*) $x_1 = 4$; $x_2=0$; $x_3=-1$
9. (*a*) No solution
   (*b*) $x=0$; $y=3$; $z=-1$
   (*c*) No solution

Exercise 18.

1. (*a*) $Y_t=8\ (5)^t$
   (*b*) $Y_t=2\ (-3)^t$
   (*c*) $Y_t=4\ (\frac{1}{2})^t$
   (*d*) $Y_t=12\ (-\frac{1}{4})^t$
2. (*a*) Explosive
   (*b*) Explosive and oscillating
   (*c*) Damped
   (*d*) Damped and oscillating

4. (*a*) $Y_t = 2\,(4)^t + 3$
   (*b*) $Y_t = 11\,(6)^t + 2$
   (*c*) $Y_t = -13\,(-3)^t + 5$
6. (*a*) $Y_t = 3^t + 1$
   (*b*) $Y_t = 2\,(5)^t + 6\,(-2)^t$
   (*c*) $Y_t = 55\,(6)^t + 2\,(2)^t$
   (*d*) $Y_t = 3\,(2)^t + 5\,(-2)^t$
7. (*a*) $Y_t = 3\,(3)^t + 5\,(-1)^t - 2$
   (*b*) $Y_t = 3\,(6)^t + 2\,(-3)^t + 2$
   (*c*) $Y_t = 3\,(4)^t + 2\,(\frac{5}{4})^t + 4$
   (*d*) $Y^t = 3\,(2)^t - 5(-2)^t - 30$

LOGARITHMS

	0	1	2	3	4	5	6	7	8	9	1	2	3	4	5	6	7	8	9
10	·0000	0043	0086	0128	0170	0212	0253	0294	0334	0374	4	8	12	17	21	25	29	33	37
11	·0414	0453	0492	0531	0569	0607	0645	0682	0719	0755	4	8	11	15	19	23	26	30	34
12	·0792	0828	0864	0899	0934	0969	1004	1038	1072	1106	3	7	10	14	17	21	24	28	31
13	·1139	1173	1206	1239	1271	1303	1335	1367	1399	1430	3	6	10	13	16	19	23	26	29
14	·1461	1492	1523	1553	1584	1614	1644	1673	1703	1732	3	6	9	12	15	18	21	24	27
15	·1761	1790	1818	1847	1875	1903	1931	1959	1987	2014	3	6	8	11	14	17	20	22	25
16	·2041	2068	2095	2122	2148	2175	2201	2227	2253	2279	3	5	8	11	13	16	18	21	24
17	·2304	2330	2355	2380	2405	2430	2455	2480	2504	2529	2	5	7	10	12	15	17	20	22
18	·2553	2577	2601	2625	2648	2672	2695	2718	2742	2765	2	5	7	9	12	14	16	19	21
19	·2788	2810	2833	2856	2878	2900	2923	2945	2967	2989	2	4	7	9	11	13	16	18	20
20	·3010	3032	3054	3075	3096	3118	3139	3160	3181	3201	2	4	6	8	11	13	15	17	19
21	·3222	3243	3263	3284	3304	3324	3345	3365	3385	3404	2	4	6	8	10	12	14	16	18
22	·3424	3444	3464	3483	3502	3522	3541	3560	3579	3598	2	4	6	8	10	12	14	15	17
23	·3617	3636	3655	3674	3692	3711	3729	3747	3766	3784	2	4	6	7	9	11	13	15	17
24	·3802	3820	3838	3856	3874	3892	3909	3927	3945	3962	2	4	5	7	9	11	12	14	16
25	·3979	3997	4014	4031	4048	4065	4082	4099	4116	4133	2	3	5	7	9	10	12	14	15
26	·4150	4166	4183	4200	4216	4232	4249	4265	4281	4298	2	3	5	7	8	10	11	13	15
27	·4314	4330	4346	4362	4378	4393	4409	4425	4440	4456	2	3	5	6	8	9	11	13	14
28	·4472	4487	4502	4518	4533	4548	4564	4579	4594	4609	2	3	5	6	8	9	11	12	14
29	·4624	4639	4654	4669	4683	4698	4713	4728	4742	4757	1	3	4	6	7	9	10	12	13
30	·4771	4786	4800	4814	4829	4843	4857	4871	4886	4900	1	3	4	6	7	9	10	11	13
31	·4914	4928	4942	4955	4969	4983	4997	5011	5024	5038	1	3	4	6	7	8	10	11	12
32	·5051	5065	5079	5092	5105	5119	5132	5145	5159	5172	1	3	4	5	7	8	9	11	12
33	·5185	5198	5211	5224	5237	5250	5263	5276	5289	5302	1	3	4	5	6	8	9	10	12
34	·5315	5328	5340	5353	5366	5378	5391	5403	5416	5428	1	3	4	5	6	8	9	10	11
35	·5441	5453	5465	5478	5490	5502	5514	5527	5539	5551	1	2	4	5	6	7	9	10	11
36	·5563	5575	5587	5599	5611	5623	5635	5647	5658	5670	1	2	4	5	6	7	8	10	11
37	·5682	5694	5705	5717	5729	5740	5752	5763	5775	5786	1	2	3	5	6	7	8	9	10
38	·5798	5809	5821	5832	5843	5855	5866	5877	5888	5899	1	2	3	5	6	7	8	9	10
39	·5911	5922	5933	5944	5955	5966	5977	5988	5999	6010	1	2	3	4	5	7	8	9	10
40	·6021	6031	6042	6053	6064	6075	6085	6096	6107	6117	1	2	3	4	5	6	8	9	10
41	·6128	6138	6149	6160	6170	6180	6191	6201	6212	6222	1	2	3	4	5	6	7	8	9
42	·6232	6243	6253	6263	6274	6284	6294	6304	6314	6325	1	2	3	4	5	6	7	8	9
43	·6335	6345	6355	6365	6375	6385	6395	6405	6415	6425	1	2	3	4	5	6	7	8	9
44	·6435	6444	6454	6464	6474	6484	6493	6503	6513	6522	1	2	3	4	5	6	7	8	9
45	·6532	6542	6551	6561	6571	6580	6590	6599	6609	6618	1	2	3	4	5	6	7	8	9
46	·6628	6637	6646	6656	6665	6675	6684	6693	6702	6712	1	2	3	4	5	6	7	7	8
47	·6721	6730	6739	6749	6758	6767	6776	6785	6794	6803	1	2	3	4	5	5	6	7	8
48	·6812	6821	6830	6839	6848	6857	6866	6875	6884	6893	1	2	3	4	4	5	6	7	8
49	·6902	6911	6920	6928	6937	6946	6955	6964	6972	6981	1	2	3	4	4	5	6	7	8
50	·6990	6998	7007	7016	7024	7033	7042	7050	7059	7067	1	2	3	3	4	5	6	7	8
51	·7076	7084	7093	7101	7110	7118	7126	7135	7143	7152	1	2	3	3	4	5	6	7	8
52	·7160	7168	7177	7185	7193	7202	7210	7218	7226	7235	1	2	2	3	4	5	6	7	7
53	·7243	7251	7259	7267	7275	7284	7292	7300	7308	7316	1	2	2	3	4	5	6	6	7
54	·7324	7332	7340	7348	7356	7364	7372	7380	7388	7396	1	2	2	3	4	5	6	6	7
	0	1	2	3	4	5	6	7	8	9	1	2	3	4	5	6	7	8	9

LOGARITHMS

	0	1	2	3	4	5	6	7	8	9	1	2	3	4	5	6	7	8	9
55	·7404	7412	7419	7427	7435	7443	7451	7459	7466	7474	1	2	2	3	4	5	5	6	7
56	·7482	7490	7497	7505	7513	7520	7528	7536	7543	7551	1	2	2	3	4	5	5	6	7
57	·7559	7566	7574	7582	7589	7597	7604	7612	7619	7627	1	2	2	3	4	5	5	6	7
58	·7634	7642	7649	7657	7664	7672	7679	7686	7694	7701	1	1	2	3	4	4	5	6	7
59	·7709	7716	7723	7731	7738	7745	7752	7760	7767	7774	1	1	2	3	4	4	5	6	7
60	·7782	7789	7796	7803	7810	7818	7825	7832	7839	7846	1	1	2	3	4	4	5	6	6
61	·7853	7860	7868	7875	7882	7889	7896	7903	7910	7917	1	1	2	3	4	4	5	6	6
62	·7924	7931	7938	7945	7952	7959	7966	7973	7980	7987	1	1	2	3	3	4	5	6	6
63	·7993	8000	8007	8014	8021	8028	8035	8041	8048	8055	1	1	2	3	3	4	5	5	6
64	·8062	8069	8075	8082	8089	8096	8102	8109	8116	8122	1	1	2	3	3	4	5	5	6
65	·8129	8136	8142	8149	8156	8162	8169	8176	8182	8189	1	1	2	3	3	4	5	5	6
66	·8195	8202	8209	8215	8222	8228	8235	8241	8248	8254	1	1	2	3	3	4	5	5	6
67	·8261	8267	8274	8280	8287	8293	8299	8306	8312	8319	1	1	2	3	3	4	5	5	6
68	·8325	8331	8338	8344	8351	8357	8363	8370	8376	8382	1	1	2	3	3	4	4	5	6
69	·8388	8395	8401	8407	8414	8420	8426	8432	8439	8445	1	1	2	2	3	4	4	5	6
70	·8451	8457	8463	8470	8476	8482	8488	8494	8500	8506	1	1	2	2	3	4	4	5	6
71	·8513	8519	8525	8531	8537	8543	8549	8555	8561	8567	1	1	2	2	3	4	4	5	5
72	·8573	8579	8585	8591	8597	8603	8609	8615	8621	8627	1	1	2	2	3	4	4	5	5
73	·8633	8639	8645	8651	8657	8663	8669	8675	8681	8686	1	1	2	2	3	4	4	5	5
74	·8692	8698	8704	8710	8716	8722	8727	8733	8739	8745	1	1	2	2	3	4	4	5	5
75	·8751	8756	8762	8768	8774	8779	8785	8791	8797	8802	1	1	2	2	3	3	4	5	5
76	·8808	8814	8820	8825	8831	8837	8842	8848	8854	8859	1	1	2	2	3	3	4	5	5
77	·8865	8871	8876	8882	8887	8893	8899	8904	8910	8915	1	1	2	2	3	3	4	4	5
78	·8921	8927	8932	8938	8943	8949	8954	8960	8965	8971	1	1	2	2	3	3	4	4	5
79	·8976	8982	8987	8993	8998	9004	9009	9015	9020	9025	1	1	2	2	3	3	4	4	5
80	·9031	9036	9042	9047	9053	9058	9063	9069	9074	9079	1	1	2	2	3	3	4	4	5
81	·9085	9090	9096	9101	9106	9112	9117	9122	9128	9133	1	1	2	2	3	3	4	4	5
82	·9138	9143	9149	9154	9159	9165	9170	9175	9180	9186	1	1	2	2	3	3	4	4	5
83	·9191	9196	9201	9206	9212	9217	9222	9227	9232	9238	1	1	2	2	3	3	4	4	5
84	·9243	9248	9253	9258	9263	9269	9274	9279	9284	9289	1	1	2	2	3	3	4	4	5
85	·9294	9299	9304	9309	9315	9320	9325	9330	9335	9340	1	1	2	2	3	3	4	4	5
86	·9345	9350	9355	9360	9365	9370	9375	9380	9385	9390	1	1	1	2	3	3	4	4	5
87	·9395	9400	9405	9410	9415	9420	9425	9430	9435	9440	0	1	1	2	2	3	3	4	4
88	·9445	9450	9455	9460	9465	9469	9474	9479	9484	9489	0	1	1	2	2	3	3	4	4
89	·9494	9499	9504	9509	9513	9518	9523	9528	9533	9538	0	1	1	2	2	3	3	4	4
90	·9542	9547	9552	9557	9562	9566	9571	9576	9581	9586	0	1	1	2	2	3	3	4	4
91	·9590	9595	9600	9605	9609	9614	9619	9624	9628	9633	0	1	1	2	2	3	3	4	4
92	·9638	9643	9647	9652	9657	9661	9666	9671	9675	9680	0	1	1	2	2	3	3	4	4
93	·9685	9689	9694	9699	9703	9708	9713	9717	9722	9727	0	1	1	2	2	3	3	4	4
94	·9731	9736	9741	9745	9750	9754	9759	9763	9768	9773	0	1	1	2	2	3	3	4	4
95	·9777	9782	9786	9791	9795	9800	9805	9809	9814	9818	0	1	1	2	2	3	3	4	4
96	·9823	9827	9832	9836	9841	9845	9850	9854	9859	9863	0	1	1	2	2	3	3	4	4
97	·9868	9872	9877	9881	9886	9890	9894	9899	9903	9908	0	1	1	2	2	3	3	4	4
98	·9912	9917	9921	9926	9930	9934	9939	9943	9948	9952	0	1	1	2	2	3	3	4	4
99	·9956	9961	9965	9969	9974	9978	9983	9987	9991	9996	0	1	1	2	2	3	3	3	4
	0	1	2	3	4	5	6	7	8	9	1	2	3	4	5	6	7	8	9

ANTI-LOGARITHMS

	0	1	2	3	4	5	6	7	8	9	1	2	3	4	5	6	7	8	9
·00	1000	1002	1005	1007	1009	1012	1014	1016	1019	1021	0	0	1	1	1	1	2	2	2
·01	1023	1026	1028	1030	1033	1035	1038	1040	1042	1045	0	0	1	1	1	1	2	2	2
·02	1047	1050	1052	1054	1057	1059	1062	1064	1067	1069	0	0	1	1	1	1	2	2	2
·03	1072	1074	1076	1079	1081	1084	1086	1089	1091	1094	0	0	1	1	1	1	2	2	2
·04	1096	1099	1102	1104	1107	1109	1112	1114	1117	1119	0	1	1	1	1	2	2	2	2
·05	1122	1125	1127	1130	1132	1135	1138	1140	1143	1146	0	1	1	1	1	2	2	2	2
·06	1148	1151	1153	1156	1159	1161	1164	1167	1169	1172	0	1	1	1	1	2	2	2	2
·07	1175	1178	1180	1183	1186	1189	1191	1194	1197	1199	0	1	1	1	1	2	2	2	2
·08	1202	1205	1208	1211	1213	1216	1219	1222	1225	1227	0	1	1	1	1	2	2	2	3
·09	1230	1233	1236	1239	1242	1245	1247	1250	1253	1256	0	1	1	1	1	2	2	2	3
·10	1259	1262	1265	1268	1271	1274	1276	1279	1282	1285	0	1	1	1	1	2	2	2	3
·11	1288	1291	1294	1297	1300	1303	1306	1309	1312	1315	0	1	1	1	2	2	2	2	3
·12	1318	1321	1324	1327	1330	1334	1337	1340	1343	1346	0	1	1	1	2	2	2	2	3
·13	1349	1352	1355	1358	1361	1365	1368	1371	1374	1377	0	1	1	1	2	2	2	3	3
·14	1380	1384	1387	1390	1393	1396	1400	1403	1406	1409	0	1	1	1	2	2	2	3	3
·15	1413	1416	1419	1422	1426	1429	1432	1435	1439	1442	0	1	1	1	2	2	2	3	3
·16	1445	1449	1452	1455	1459	1462	1466	1469	1472	1476	0	1	1	1	2	2	2	3	3
·17	1479	1483	1486	1489	1493	1496	1500	1503	1507	1510	0	1	1	1	2	2	2	3	3
·18	1514	1517	1521	1524	1528	1531	1535	1538	1542	1545	0	1	1	1	2	2	2	3	3
·19	1549	1552	1556	1560	1563	1567	1570	1574	1578	1581	0	1	1	1	2	2	3	3	3
·20	1585	1589	1592	1596	1600	1603	1607	1611	1614	1618	0	1	1	1	2	2	3	3	3
·21	1622	1626	1629	1633	1637	1641	1644	1648	1652	1656	0	1	1	2	2	2	3	3	3
·22	1660	1663	1667	1671	1675	1679	1683	1687	1690	1694	0	1	1	2	2	2	3	3	3
·23	1698	1702	1706	1710	1714	1718	1722	1726	1730	1734	0	1	1	2	2	2	3	3	4
·24	1738	1742	1746	1750	1754	1758	1762	1766	1770	1774	0	1	1	2	2	2	3	3	4
·25	1778	1782	1786	1791	1795	1799	1803	1807	1811	1816	0	1	1	2	2	2	3	3	4
·26	1820	1824	1828	1832	1837	1841	1845	1849	1854	1858	0	1	1	2	2	3	3	3	4
·27	1862	1866	1871	1875	1879	1884	1888	1892	1897	1901	0	1	1	2	2	3	3	3	4
·28	1905	1910	1914	1919	1923	1928	1932	1936	1941	1945	0	1	1	2	2	3	3	4	4
·29	1950	1954	1959	1963	1968	1972	1977	1982	1986	1991	0	1	1	2	2	3	3	4	4
·30	1995	2000	2004	2009	2014	2018	2023	2028	2032	2037	0	1	1	2	2	3	3	4	4
·31	2042	2046	2051	2056	2061	2065	2070	2075	2080	2084	0	1	1	2	2	3	3	4	4
·32	2089	2094	2099	2104	2109	2113	2118	2123	2128	2133	0	1	1	2	2	3	3	4	4
·33	2138	2143	2148	2153	2158	2163	2168	2173	2178	2183	0	1	1	2	2	3	3	4	4
·34	2188	2193	2198	2203	2208	2213	2218	2223	2228	2234	1	1	2	2	3	3	4	4	5
·35	2239	2244	2249	2254	2259	2265	2270	2275	2280	2286	1	1	2	2	3	3	4	4	5
·36	2291	2296	2301	2307	2312	2317	2323	2328	2333	2339	1	1	2	2	3	3	4	4	5
·37	2344	2350	2355	2360	2366	2371	2377	2382	2388	2393	1	1	2	2	3	3	4	4	5
·38	2399	2404	2410	2415	2421	2427	2432	2438	2443	2449	1	1	2	2	3	3	4	4	5
·39	2455	2460	2466	2472	2477	2483	2489	2495	2500	2506	1	1	2	2	3	3	4	5	5
·40	2512	2518	2523	2529	2535	2541	2547	2553	2559	2564	1	1	2	2	3	4	4	5	5
·41	2570	2576	2582	2588	2594	2600	2606	2612	2618	2624	1	1	2	2	3	4	4	5	5
·42	2630	2636	2642	2649	2655	2661	2667	2673	2679	2685	1	1	2	2	3	4	4	5	6
·43	2692	2698	2704	2710	2716	2723	2729	2735	2742	2748	1	1	2	3	3	4	4	5	6
·44	2754	2761	2767	2773	2780	2786	2793	2799	2805	2812	1	1	2	3	3	4	4	5	6
·45	2818	2825	2831	2838	2844	2851	2858	2864	2871	2877	1	1	2	3	3	4	5	5	6
·46	2884	2891	2897	2904	2911	2917	2924	2931	2938	2944	1	1	2	3	3	4	5	5	6
·47	2951	2958	2965	2972	2979	2985	2992	2999	3006	3013	1	1	2	3	3	4	5	5	6
·48	3020	3027	3034	3041	3048	3055	3062	3069	3076	3083	1	1	2	3	4	4	5	6	6
·49	3090	3097	3105	3112	3119	3126	3133	3141	3148	3155	1	1	2	3	4	4	5	6	6
	0	1	2	3	4	5	6	7	8	9	1	2	3	4	5	6	7	8	9

ANTI-LOGARITHMS

	0	1	2	3	4	5	6	7	8	9	1	2	3	4	5	6	7	8	9
·50	3162	3170	3177	3184	3192	3199	3206	3214	3221	3228	1	1	2	3	4	4	5	6	7
·51	3236	3243	3251	3258	3266	3273	3281	3289	3296	3304	1	2	2	3	4	5	5	6	7
·52	3311	3319	3327	3334	3342	3350	3357	3365	3373	3381	1	2	2	3	4	5	5	6	7
·53	3388	3396	3404	3412	3420	3428	3436	3443	3451	3459	1	2	2	3	4	5	6	6	7
·54	3467	3475	3483	3491	3499	3508	3516	3524	3532	3540	1	2	2	3	4	5	6	6	7
·55	3548	3556	3565	3573	3581	3589	3597	3606	3614	3622	1	2	2	3	4	5	6	7	7
·56	3631	3639	3648	3656	3664	3673	3681	3690	3698	3707	1	2	3	3	4	5	6	7	8
·57	3715	3724	3733	3741	3750	3758	3767	3776	3784	3793	1	2	3	3	4	5	6	7	8
·58	3802	3811	3819	3828	3837	3846	3855	3864	3873	3882	1	2	3	4	4	5	6	7	8
·59	3890	3899	3908	3917	3926	3936	3945	3954	3963	3972	1	2	3	4	5	5	6	7	8
·60	3981	3990	3999	4009	4018	4027	4036	4046	4055	4064	1	2	3	4	5	6	6	7	8
·61	4074	4083	4093	4102	4111	4121	4130	4140	4150	4159	1	2	3	4	5	6	7	8	9
·62	4169	4178	4188	4198	4207	4217	4227	4236	4246	4256	1	2	3	4	5	6	7	8	9
·63	4266	4276	4285	4295	4305	4315	4325	4335	4345	4355	1	2	3	4	5	6	7	8	9
·64	4365	4375	4385	4395	4406	4416	4426	4436	4446	4457	1	2	3	4	5	6	7	8	9
·65	4467	4477	4487	4498	4508	4519	4529	4539	4550	4560	1	2	3	4	5	6	7	8	9
·66	4571	4581	4592	4603	4613	4624	4634	4645	4656	4667	1	2	3	4	5	6	7	9	10
·67	4677	4688	4699	4710	4721	4732	4742	4753	4764	4775	1	2	3	4	5	7	8	9	10
·68	4786	4797	4808	4819	4831	4842	4853	4864	4875	4887	1	2	3	4	6	7	8	9	10
·69	4898	4909	4920	4932	4943	4955	4966	4977	4989	5000	1	2	3	5	6	7	8	9	10
·70	5012	5023	5035	5047	5058	5070	5082	5093	5105	5117	1	2	4	5	6	7	8	9	11
·71	5129	5140	5152	5164	5176	5188	5200	5212	5224	5236	1	2	4	5	6	7	8	10	11
·72	5248	5260	5272	5284	5297	5309	5321	5333	5346	5358	1	2	4	5	6	7	9	10	11
·73	5370	5383	5395	5408	5420	5433	5445	5458	5470	5483	1	3	4	5	6	8	9	10	11
·74	5495	5508	5521	5534	5546	5559	5572	5585	5598	5610	1	3	4	5	6	8	9	10	12
·75	5623	5636	5649	5662	5675	5689	5702	5715	5728	5741	1	3	4	5	7	8	9	10	12
·76	5754	5768	5781	5794	5808	5821	5834	5848	5861	5875	1	3	4	5	7	8	9	11	12
·77	5888	5902	5916	5929	5943	5957	5970	5984	5998	6012	1	3	4	5	7	8	10	11	12
·78	6026	6039	6053	6067	6081	6095	6109	6124	6138	6152	1	3	4	6	7	8	10	11	13
·79	6166	6180	6194	6209	6223	6237	6252	6266	6281	6295	1	3	4	6	7	9	10	11	13
·80	6310	6324	6339	6353	6368	6383	6397	6412	6427	6442	1	3	4	6	7	9	10	12	13
·81	6457	6471	6486	6501	6516	6531	6546	6561	6577	6592	2	3	5	6	8	9	11	12	14
·82	6607	6622	6637	6653	6668	6683	6699	6714	6730	6745	2	3	5	6	8	9	11	12	14
·83	6761	6776	6792	6808	6823	6839	6855	6871	6887	6902	2	3	5	6	8	9	11	13	14
·84	6918	6934	6950	6966	6982	6998	7015	7031	7047	7063	2	3	5	6	8	10	11	13	15
·85	7079	7096	7112	7129	7145	7161	7178	7194	7211	7228	2	3	5	7	8	10	12	13	15
·86	7244	7261	7278	7295	7311	7328	7345	7362	7379	7396	2	3	5	7	8	10	12	13	15
·87	7413	7430	7447	7464	7482	7499	7516	7534	7551	7568	2	3	5	7	9	10	12	14	16
·88	7586	7603	7621	7638	7656	7674	7691	7709	7727	7745	2	4	5	7	9	11	12	14	16
·89	7762	7780	7798	7816	7834	7852	7870	7889	7907	7925	2	4	5	7	9	11	13	14	16
·90	7943	7962	7980	7998	8017	8035	8054	8072	8091	8110	2	4	6	7	9	11	13	15	17
·91	8128	8147	8166	8185	8204	8222	8241	8260	8279	8299	2	4	6	8	9	11	13	15	17
·92	8318	8337	8356	8375	8395	8414	8433	8453	8472	8492	2	4	6	8	10	12	14	15	17
·93	8511	8531	8551	8570	8590	8610	8630	8650	8670	8690	2	4	6	8	10	12	14	16	18
·94	8710	8730	8750	8770	8790	8810	8831	8851	8872	8892	2	4	6	8	10	12	14	16	18
·95	8913	8933	8954	8974	8995	9016	9036	9057	9078	9099	2	4	6	8	10	12	15	17	19
·96	9120	9141	9162	9183	9204	9226	9247	9268	9290	9311	2	4	6	8	11	13	15	17	19
·97	9333	9354	9376	9397	9419	9441	9462	9484	9506	9528	2	4	7	9	11	13	15	17	20
·98	9550	9572	9594	9616	9638	9661	9683	9705	9727	9750	2	4	7	9	11	13	16	18	20
·99	9772	9795	9817	9840	9863	9886	9908	9931	9954	9977	2	5	7	9	11	14	16	18	20
	0	1	2	3	4	5	6	7	8	9	1	2	3	4	5	6	7	8	9

SQUARES

	0	1	2	3	4	5	6	7	8	9	1	2	3	4	5	6	7	8	9
10	1000	1020	1040	1061	1082	1103	1124	1145	1166	1188	2	4	6	8	10	13	15	17	19
11	1210	1232	1254	1277	1300	1323	1346	1369	1392	1416	2	5	7	9	11	14	16	18	21
12	1440	1464	1488	1513	1538	1563	1588	1613	1638	1664	2	5	7	10	12	15	17	20	22
13	1690	1716	1742	1769	1796	1823	1850	1877	1904	1932	3	5	8	11	13	16	19	22	24
14	1960	1988	2016	2045	2074	2103	2132	2161	2190	2220	3	6	9	12	14	17	20	23	26
15	2250	2280	2310	2341	2372	2403	2434	2465	2496	2528	3	6	9	12	15	19	22	25	28
16	2560	2592	2624	2657	2690	2723	2756	2789	2822	2856	3	7	10	13	16	20	23	26	30
17	2890	2924	2958	2993	3028	3063	3098	3133	3168	3204	3	7	10	14	17	21	24	28	31
18	3240	3276	3312	3349	3386	3423	3460	3497	3534	3572	4	7	11	15	18	22	26	30	33
19	3610	3648	3686	3725	3764	3803	3842	3881	3920	3960	4	8	12	16	19	23	27	31	35
20	4000	4040	4080	4121	4162	4203	4244	4285	4326	4368	4	8	12	16	20	25	29	33	37
21	4410	4452	4494	4537	4580	4623	4666	4709	4752	4796	4	9	13	17	21	26	30	34	39
22	4840	4884	4928	4973	5018	5063	5108	5153	5198	5244	4	9	13	18	22	27	31	36	40
23	5290	5336	5382	5429	5476	5523	5570	5617	5664	5712	5	9	14	19	23	28	33	38	42
24	5760	5808	5856	5905	5954	6003	6052	6101	6150	6200	5	10	15	20	24	29	34	39	44
25	6250	6300	6350	6401	6452	6503	6554	6605	6656	6708	5	10	15	20	25	31	36	41	46
26	6760	6812	6864	6917	6970	7023	7076	7129	7182	7236	5	11	16	21	26	32	37	42	48
27	7290	7344	7398	7453	7508	7563	7618	7673	7728	7784	5	11	16	22	28	33	38	44	49
28	7840	7896	7952	8009	8066	8123	8180	8237	8294	8352	6	11	17	23	28	34	40	46	51
29	8410	8468	8526	8585	8644	8703	8762	8821	8880	8940	6	12	18	24	30	35	41	47	53
30	9000	9060	9120	9181	9242	9303	9364	9425	9486	9548	6	12	18	24	31	37	43	49	55
31	9610	9672	9734	9797	9860	9923	9986				6	13	19	25	31	38	44	50	56
31								1005	1011	1018	1	1	2	3	3	4	5	5	6
32	1024	1030	1037	1043	1050	1056	1063	1069	1076	1082	1	1	2	3	3	4	5	5	6
33	1089	1096	1102	1109	1116	1122	1129	1136	1142	1149	1	1	2	3	3	4	5	5	6
34	1156	1163	1170	1176	1183	1190	1197	1204	1211	1218	1	1	2	3	3	4	5	6	6
35	1225	1232	1239	1246	1253	1260	1267	1274	1282	1289	1	1	2	3	4	4	5	6	6
36	1296	1303	1310	1318	1325	1332	1340	1347	1354	1362	1	1	2	3	4	4	5	6	7
37	1369	1376	1384	1391	1399	1406	1414	1421	1429	1436	1	2	2	3	4	5	5	6	7
38	1444	1452	1459	1467	1475	1482	1490	1498	1505	1513	1	2	2	3	4	5	5	6	7
39	1521	1529	1537	1544	1552	1560	1568	1576	1584	1592	1	2	2	3	4	5	6	6	7
40	1600	1608	1616	1624	1632	1640	1648	1656	1665	1673	1	2	2	3	4	5	6	6	7
41	1681	1689	1697	1706	1714	1722	1731	1739	1747	1756	1	2	2	3	4	5	6	7	7
42	1764	1772	1781	1789	1798	1806	1815	1823	1832	1840	1	2	3	3	4	5	6	7	8
43	1849	1858	1866	1875	1884	1892	1901	1910	1918	1927	1	2	3	3	4	5	6	7	8
44	1936	1945	1954	1962	1971	1980	1989	1998	2007	2016	1	2	3	4	5	5	6	7	8
45	2025	2034	2043	2052	2061	2070	2079	2088	2098	2107	1	2	3	4	5	5	6	7	8
46	2116	2125	2134	2144	2153	2162	2172	2181	2190	2200	1	2	3	4	5	6	7	7	8
47	2209	2218	2228	2237	2247	2256	2266	2275	2285	2294	1	2	3	4	5	6	7	8	9
48	2304	2314	2323	2333	2343	2352	2362	2372	2381	2391	1	2	3	4	5	6	7	8	9
49	2401	2411	2421	2430	2440	2450	2460	2470	2480	2490	1	2	3	4	5	6	7	8	9
50	2500	2510	2520	2530	2540	2550	2560	2570	2581	2591	1	2	3	4	5	6	7	8	9
51	2601	2611	2621	2632	2642	2652	2663	2673	2683	2694	1	2	3	4	5	6	7	8	9
52	2704	2714	2725	2735	2746	2756	2767	2777	2788	2798	1	2	3	4	5	6	7	8	9
53	2809	2820	2830	2841	2852	2862	2873	2884	2894	2905	1	2	3	4	5	6	7	9	10
54	2916	2927	2938	2948	2959	2970	2981	2992	3003	3014	1	2	3	4	6	7	8	9	10
	0	1	2	3	4	5	6	7	8	9	1	2	3	4	5	6	7	8	9

The position of the decimal point must be determined by inspection.

SQUARES

	0	1	2	3	4	5	6	7	8	9	1	2	3	4	5	6	7	8	9
55	3025	3036	3047	3058	3069	3080	3091	3102	3114	3125	1	2	3	4	6	7	8	9	10
56	3136	3147	3158	3170	3181	3192	3204	3215	3226	3238	1	2	3	5	6	7	8	9	10
57	3249	3260	3272	3283	3295	3306	3318	3329	3341	3352	1	2	3	5	6	7	8	9	10
58	3364	3376	3387	3399	3411	3422	3434	3446	3457	3469	1	2	4	5	6	7	8	9	11
59	3481	3493	3505	3516	3528	3540	3552	3564	3576	3588	1	2	4	5	6	7	8	10	11
60	3600	3612	3624	3636	3648	3660	3672	3684	3697	3709	1	2	4	5	6	7	8	10	11
61	3721	3733	3745	3758	3770	3782	3795	3807	3819	3832	1	2	4	5	6	7	9	10	11
62	3844	3856	3869	3881	3894	3906	3919	3931	3944	3956	1	3	4	5	6	8	9	10	11
63	3969	3982	3994	4007	4020	4032	4045	4058	4070	4083	1	3	4	5	6	8	9	10	11
64	4096	4109	4122	4134	4147	4160	4173	4186	4199	4212	1	3	4	5	6	8	9	10	12
65	4225	4238	4251	4264	4277	4290	4303	4316	4330	4343	1	3	4	5	7	8	9	10	12
66	4356	4369	4382	4396	4409	4422	4436	4449	4462	4476	1	3	4	5	7	8	9	11	12
67	4489	4502	4516	4529	4543	4556	4570	4583	4597	4610	1	3	4	5	7	8	9	11	12
68	4624	4638	4651	4665	4679	4692	4706	4720	4733	4747	1	3	4	5	7	8	10	11	12
69	4761	4775	4789	4802	4816	4830	4844	4858	4872	4886	1	3	4	6	7	8	10	11	13
70	4900	4914	4928	4942	4956	4970	4984	4998	5013	5027	1	3	4	6	7	8	10	11	13
71	5041	5055	5069	5084	5098	5112	5127	5141	5155	5170	1	3	4	6	7	9	10	11	13
72	5184	5198	5213	5227	5242	5256	5271	5285	5300	5314	1	3	4	6	7	9	10	11	13
73	5329	5344	5358	5373	5388	5402	5417	5432	5446	5461	1	3	4	6	7	9	10	12	13
74	5476	5491	5506	5520	5535	5550	5565	5580	5595	5610	1	3	4	6	7	9	10	12	13
75	5625	5640	5655	5670	5685	5700	5715	5730	5746	5761	2	3	5	6	8	9	11	12	14
76	5776	5791	5806	5822	5837	5852	5868	5883	5898	5914	2	3	5	6	8	9	11	12	14
77	5929	5944	5960	5975	5991	6006	6022	6037	6053	6068	2	3	5	6	8	9	11	12	14
78	6084	6100	6115	6131	6147	6162	6178	6194	6209	6225	2	3	5	6	8	9	11	13	14
79	6241	6257	6273	6288	6304	6320	6336	6352	6368	6384	2	3	5	6	8	10	11	13	14
80	6400	6416	6432	6448	6464	6480	6496	6512	6529	6545	2	3	5	6	8	10	11	13	14
81	6561	6577	6593	6610	6626	6642	6659	6675	6691	6708	2	3	5	7	8	10	11	13	15
82	6724	6740	6757	6773	6790	6806	6823	6839	6856	6872	2	3	5	7	8	10	11	13	15
83	6889	6906	6922	6939	6956	6972	6989	7006	7022	7039	2	3	5	7	8	10	12	13	15
84	7056	7073	7090	7106	7123	7140	7157	7174	7191	7208	2	3	5	7	8	10	12	14	15
85	7225	7242	7259	7276	7293	7310	7327	7344	7362	7379	2	3	5	7	9	10	12	14	15
86	7396	7413	7430	7448	7465	7482	7500	7517	7534	7552	2	3	5	7	9	10	12	14	16
87	7569	7586	7604	7621	7639	7656	7674	7691	7709	7726	2	4	5	7	9	11	12	14	16
88	7744	7762	7779	7797	7815	7832	7850	7868	7885	7903	2	4	5	7	9	11	12	14	16
89	7921	7939	7957	7974	7992	8010	8028	8046	8064	8082	2	4	5	7	9	11	13	14	16
90	8100	8118	8136	8154	8172	8190	8208	8226	8245	8263	2	4	5	7	9	11	13	14	16
91	8281	8299	8317	8336	8354	8372	8391	8409	8427	8446	2	4	5	7	9	11	13	15	16
92	8464	8482	8501	8519	8538	8556	8575	8593	8612	8630	2	4	6	7	9	11	13	15	17
93	8649	8668	8686	8705	8724	8742	8761	8780	8798	8817	2	4	6	7	9	11	13	15	17
94	8836	8855	8874	8892	8911	8930	8949	8968	8987	9006	2	4	6	8	9	11	13	15	17
95	9025	9044	9063	9082	9101	9120	9139	9158	9178	9197	2	4	6	8	10	11	13	15	17
96	9216	9235	9254	9274	9293	9312	9332	9351	9370	9390	2	4	6	8	10	12	14	15	17
97	9409	9428	9448	9467	9487	9506	9526	9545	9565	9584	2	4	6	8	10	12	14	16	18
98	9604	9624	9643	9663	9683	9702	9722	9742	9761	9781	2	4	6	8	10	12	14	16	18
99	9801	9821	9841	9860	9880	9900	9920	9940	9960	9980	2	4	6	8	10	12	14	16	18
	0	1	2	3	4	5	6	7	8	9	1	2	3	4	5	6	7	8	9

The position of the decimal point must be determined by inspection.

SQUARE ROOTS

	0	1	2	3	4	5	6	7	8	9	1	2	3	4	5	6	7	8	9
10	1000 3162	1005 3178	1010 3194	1015 3209	1020 3225	1025 3240	1030 3256	1034 3271	1039 3286	1044 3302	0 2	1 3	1 5	2 6	2 8	3 9	3 11	4 12	4 14
11	1049 3317	1054 3332	1058 3347	1063 3362	1068 3376	1072 3391	1077 3406	1082 3421	1086 3435	1091 3450	0 1	1 3	1 4	2 6	2 7	3 9	3 10	4 12	4 13
12	1095 3464	1100 3479	1105 3493	1109 3507	1114 3521	1118 3536	1122 3550	1127 3564	1131 3578	1136 3592	0 1	1 3	1 4	2 6	2 7	3 8	3 10	4 11	4 13
13	1140 3606	1145 3619	1149 3633	1153 3647	1158 3661	1162 3674	1166 3688	1170 3701	1175 3715	1179 3728	0 1	1 3	1 4	2 5	2 7	3 8	3 10	3 11	4 12
14	1183 3742	1187 3755	1192 3768	1196 3782	1200 3795	1204 3808	1208 3821	1212 3834	1217 3847	1221 3860	0 1	1 3	1 4	2 5	2 7	3 8	3 9	3 11	4 12
15	1225 3873	1229 3886	1233 3899	1237 3912	1241 3924	1245 3937	1249 3950	1253 3962	1257 3975	1261 3987	0 1	1 3	1 4	2 5	2 6	3 8	3 9	3 10	4 11
16	1265 4000	1269 4012	1273 4025	1277 4037	1281 4050	1285 4062	1288 4074	1292 4087	1296 4099	1300 4111	0 1	1 2	1 4	2 5	2 6	3 7	3 9	3 10	4 11
17	1304 4123	1308 4135	1311 4147	1315 4159	1319 4171	1323 4183	1327 4195	1330 4207	1334 4219	1338 4231	0 1	1 2	1 4	2 5	2 6	2 7	3 8	3 10	3 11
18	1342 4243	1345 4254	1349 4266	1353 4278	1356 4290	1360 4301	1364 4313	1367 4324	1371 4336	1375 4347	0 1	1 2	1 3	1 5	2 6	2 7	3 8	3 9	3 10
19	1378 4359	1382 4370	1386 4382	1389 4393	1393 4405	1396 4416	1400 4427	1404 4438	1407 4450	1411 4461	0 1	1 2	1 3	1 5	2 6	2 7	3 8	3 9	3 10
20	1414 4472	1418 4483	1421 4494	1425 4506	1428 4517	1432 4528	1435 4539	1439 4550	1442 4561	1446 4572	0 1	1 2	1 3	1 4	2 5	2 7	2 8	3 9	3 10
21	1449 4583	1453 4593	1456 4604	1459 4615	1463 4626	1466 4637	1470 4648	1473 4658	1476 4669	1480 4680	0 1	1 2	1 3	1 4	2 5	2 6	2 8	3 9	3 10
22	1483 4690	1487 4701	1490 4712	1493 4722	1497 4733	1500 4743	1503 4754	1507 4764	1510 4775	1513 4785	0 1	1 2	1 3	1 4	2 5	2 6	2 7	3 8	3 9
23	1517 4796	1520 4806	1523 4817	1526 4827	1530 4837	1533 4848	1536 4858	1539 4868	1543 4879	1546 4889	0 1	1 2	1 3	1 4	2 5	2 6	2 7	3 8	3 9
24	1549 4899	1552 4909	1556 4919	1559 4930	1562 4940	1565 4950	1568 4960	1572 4970	1575 4980	1578 4990	0 1	1 2	1 3	1 4	2 5	2 6	2 7	3 8	3 9
25	1581 5000	1584 5010	1587 5020	1591 5030	1594 5040	1597 5050	1600 5060	1603 5070	1606 5079	1609 5089	0 1	1 2	1 3	1 4	2 5	2 6	2 7	3 8	3 9
26	1612 5099	1616 5109	1619 5119	1622 5128	1625 5138	1628 5148	1631 5158	1634 5167	1637 5177	1640 5187	0 1	1 2	1 3	1 4	2 5	2 6	2 7	2 8	3 9
27	1643 5196	1646 5206	1649 5215	1652 5225	1655 5235	1658 5244	1661 5254	1664 5263	1667 5273	1670 5282	0 1	1 2	1 3	1 4	2 5	2 6	2 7	2 8	3 9
28	1673 5292	1676 5301	1679 5310	1682 5320	1685 5329	1688 5339	1691 5348	1694 5357	1697 5367	1700 5376	0 1	1 2	1 3	1 4	1 5	2 6	2 7	2 7	3 8
29	1703 5385	1706 5394	1709 5404	1712 5413	1715 5422	1718 5431	1720 5441	1723 5450	1726 5459	1729 5468	0 1	1 2	1 3	1 4	1 5	2 5	2 6	2 7	3 8
30	1732 5477	1735 5486	1738 5495	1741 5505	1744 5514	1746 5523	1749 5532	1752 5541	1755 5550	1758 5559	0 1	1 2	1 3	1 4	1 4	2 5	2 6	2 7	3 8
31	1761 5568	1764 5577	1766 5586	1769 5595	1772 5604	1775 5612	1778 5621	1780 5630	1783 5639	1786 5648	0 1	1 2	1 3	1 3	1 4	2 5	2 6	2 7	3 8
32	1789 5657	1792 5666	1794 5675	1797 5683	1800 5692	1803 5701	1806 5710	1808 5718	1811 5727	1814 5736	0 1	1 2	1 3	1 3	1 4	2 5	2 6	2 7	2 8
	0	1	2	3	4	5	6	7	8	9	1	2	3	4	5	6	7	8	9

The first significant figure and the position of the decimal point must be determined by inspection.

SQUARE ROOTS

	0	1	2	3	4	5	6	7	8	9	1	2	3	4	5	6	7	8	9
33	1817 5745	1819 5753	1822 5762	1825 5771	1828 5779	1830 5788	1833 5797	1836 5805	1838 5814	1841 5822	0 1	1 2	1 3	1 3	1 4	2 5	2 6	2 7	2 8
34	1844 5831	1847 5840	1849 5848	1852 5857	1855 5865	1857 5874	1860 5882	1863 5891	1865 5899	1868 5908	0 1	1 2	1 3	1 3	1 4	2 5	2 6	2 7	2 8
35	1871 5916	1873 5925	1876 5933	1879 5941	1881 5950	1884 5958	1887 5967	1889 5975	1892 5983	1895 5992	0 1	1 2	1 2	1 3	1 4	2 5	2 6	2 7	2 8
36	1897 6000	1900 6008	1903 6017	1905 6025	1908 6033	1910 6042	1913 6050	1916 6058	1918 6066	1921 6075	0 1	1 2	1 2	1 3	1 4	2 5	2 6	2 7	2 7
37	1924 6083	1926 6091	1929 6099	1931 6107	1934 6116	1936 6124	1939 6132	1942 6140	1944 6148	1947 6156	0 1	1 2	1 2	1 3	1 4	2 5	2 6	2 7	2 7
38	1949 6164	1952 6173	1954 6181	1957 6189	1960 6197	1962 6205	1965 6213	1967 6221	1970 6229	1972 6237	0 1	1 2	1 2	1 3	1 4	2 5	2 6	2 6	2 7
39	1975 6245	1977 6253	1980 6261	1982 6269	1985 6277	1987 6285	1990 6293	1992 6301	1995 6309	1997 6317	0 1	1 2	1 2	1 3	1 4	2 5	2 6	2 6	2 7
40	2000 6325	2002 6332	2005 6340	2007 6348	2010 6356	2012 6364	2015 6372	2017 6380	2020 6387	2022 6395	0 1	0 2	1 2	1 3	1 4	1 5	2 6	2 6	2 7
41	2025 6403	2027 6411	2030 6419	2032 6427	2035 6434	2037 6442	2040 6450	2042 6458	2045 6465	2047 6473	0 1	0 2	1 2	1 3	1 4	1 5	2 5	2 6	2 7
42	2049 6481	2052 6488	2054 6496	2057 6504	2059 6512	2062 6519	2064 6527	2066 6535	2069 6542	2071 6550	0 1	0 2	1 2	1 3	1 4	1 5	2 5	2 6	2 7
43	2074 6557	2076 6565	2078 6573	2081 6580	2083 6588	2086 6595	2088 6603	2090 6611	2093 6618	2095 6626	0 1	0 2	1 2	1 3	1 4	1 5	2 5	2 6	2 7
44	2098 6633	2100 6641	2102 6648	2105 6656	2107 6663	2110 6671	2112 6678	2114 6686	2117 6693	2119 6701	0 1	0 2	1 2	1 3	1 4	1 4	2 5	2 6	2 7
45	2121 6708	2124 6716	2126 6723	2128 6731	2131 6738	2133 6745	2135 6753	2138 6760	2140 6768	2142 6775	0 1	0 1	1 2	1 3	1 4	1 4	2 5	2 6	2 7
46	2145 6782	2147 6790	2149 6797	2152 6804	2154 6812	2156 6819	2159 6826	2161 6834	2163 6841	2166 6848	0 1	0 1	1 2	1 3	1 4	1 4	2 5	2 6	2 7
47	2168 6856	2170 6863	2173 6870	2175 6877	2177 6885	2179 6892	2182 6899	2184 6907	2186 6914	2189 6921	0 1	0 1	1 2	1 3	1 4	1 4	2 5	2 6	2 7
48	2191 6928	2193 6935	2195 6943	2198 6950	2200 6957	2202 6964	2205 6971	2207 6979	2209 6986	2211 6993	0 1	0 1	1 2	1 3	1 4	1 4	2 5	2 6	2 6
49	2214 7000	2216 7007	2218 7014	2220 7021	2223 7029	2225 7036	2227 7043	2229 7050	2232 7057	2234 7064	0 1	0 1	1 2	1 3	1 4	1 4	2 5	2 6	2 6
50	2236 7071	2238 7078	2241 7085	2243 7092	2245 7099	2247 7106	2249 7113	2252 7120	2254 7127	2256 7134	0 1	0 1	1 2	1 3	1 4	1 4	2 5	2 6	2 6
51	2258 7141	2261 7148	2263 7155	2265 7162	2267 7169	2269 7176	2272 7183	2274 7190	2276 7197	2278 7204	0 1	0 1	1 2	1 3	1 4	1 4	2 5	2 6	2 6
52	2280 7211	2283 7218	2285 7225	2287 7232	2289 7239	2291 7246	2293 7253	2296 7259	2298 7266	2300 7273	0 1	0 1	1 2	1 3	1 3	1 4	2 5	2 6	2 6
53	2302 7280	2304 7287	2307 7294	2309 7301	2311 7308	2313 7314	2315 7321	2317 7328	2319 7335	2322 7342	0 1	0 1	1 2	1 3	1 3	1 4	2 5	2 5	2 6
54	2324 7348	2326 7355	2328 7362	2330 7369	2332 7376	2335 7382	2337 7389	2339 7396	2341 7403	2343 7409	0 1	0 1	1 2	1 3	1 3	1 4	1 5	2 5	2 6
	0	1	2	3	4	5	6	7	8	9	1	2	3	4	5	6	7	8	9

The first significant figure and the position of the decimal point must be determined by inspection.

SQUARE ROOTS

	0	1	2	3	4	5	6	7	8	9	1	2	3	4	5	6	7	8	9
55	2345 7416	2347 7423	2349 7430	2352 7436	2354 7443	2356 7450	2358 7457	2360 7463	2362 7470	2364 7477	0 1	0 1	1 2	1 3	1 3	1 4	1 5	2 5	2 6
56	2366 7483	2369 7490	2371 7497	2373 7503	2375 7510	2377 7517	2379 7523	2381 7530	2383 7537	2385 7543	0 1	0 1	1 2	1 3	1 3	1 4	1 5	2 5	2 6
57	2387 7550	2390 7556	2392 7563	2394 7570	2396 7576	2398 7583	2400 7589	2402 7596	2404 7603	2406 7609	0 1	0 1	1 2	1 3	1 3	1 4	1 5	2 5	2 6
58	2408 7616	2410 7622	2412 7629	2415 7635	2417 7642	2419 7649	2421 7655	2423 7662	2425 7668	2427 7675	0 1	0 1	1 2	1 3	1 3	1 4	1 5	2 5	2 6
59	2429 7681	2431 7688	2433 7694	2435 7701	2437 7707	2439 7714	2441 7720	2443 7727	2445 7733	2447 7740	0 1	0 1	1 2	1 3	1 3	1 4	1 5	2 5	2 6
60	2449 7746	2452 7752	2454 7759	2456 7765	2458 7772	2460 7778	2462 7785	2464 7791	2466 7797	2468 7804	0 1	0 1	1 2	1 3	1 3	1 4	1 4	2 5	2 6
61	2470 7810	2472 7817	2474 7823	2476 7829	2478 7836	2480 7842	2482 7849	2484 7855	2486 7861	2488 7868	0 1	0 1	1 2	1 3	1 3	1 4	1 4	2 5	2 6
62	2490 7874	2492 7880	2494 7887	2496 7893	2498 7899	2500 7906	2502 7912	2504 7918	2506 7925	2508 7931	0 1	0 1	1 2	1 3	1 3	1 4	1 4	2 5	2 6
63	2510 7937	2512 7944	2514 7950	2516 7956	2518 7962	2520 7969	2522 7975	2524 7981	2526 7987	2528 7994	0 1	0 1	1 2	1 3	1 3	1 4	1 4	2 5	2 6
64	2530 8000	2532 8006	2534 8012	2536 8019	2538 8025	2540 8031	2542 8037	2544 8044	2546 8050	2548 8056	0 1	0 1	1 2	1 2	1 3	1 4	1 4	2 5	2 6
65	2550 8062	2551 8068	2553 8075	2555 8081	2557 8087	2559 8093	2561 8099	2563 8106	2565 8112	2567 8118	0 1	0 1	1 2	1 2	1 3	1 4	1 4	2 5	2 5
66	2569 8124	2571 8130	2573 8136	2575 8142	2577 8149	2579 8155	2581 8161	2583 8167	2585 8173	2587 8179	0 1	0 1	1 2	1 2	1 3	1 4	1 4	2 5	2 5
67	2588 8185	2590 8191	2592 8198	2594 8204	2596 8210	2598 8216	2600 8222	2602 8228	2604 8234	2606 8240	0 1	0 1	1 2	1 2	1 3	1 4	1 4	2 5	2 5
68	2608 8246	2610 8252	2612 8258	2613 8264	2615 8270	2617 8276	2619 8283	2621 8289	2623 8295	2625 8301	0 1	0 1	1 2	1 2	1 3	1 4	1 4	2 5	2 5
69	2627 8307	2629 8313	2631 8319	2632 8325	2634 8331	2636 8337	2638 8343	2640 8349	2642 8355	2644 8361	0 1	0 1	1 2	1 2	1 3	1 4	1 4	2 5	2 5
70	2646 8367	2648 8373	2650 8379	2651 8385	2653 8390	2655 8396	2657 8402	2659 8408	2661 8414	2663 8420	0 1	0 1	1 2	1 2	1 3	1 4	1 4	2 5	2 5
71	2665 8426	2666 8432	2668 8438	2670 8444	2672 8450	2674 8456	2676 8462	2678 8468	2680 8473	2681 8479	0 1	0 1	1 2	1 2	1 3	1 3	1 4	1 5	2 5
72	2683 8485	2685 8491	2687 8497	2689 8503	2691 8509	2693 8515	2694 8521	2696 8526	2698 8532	2700 8538	0 1	0 1	1 2	1 2	1 3	1 3	1 4	1 5	2 5
73	2702 8544	2704 8550	2706 8556	2707 8562	2709 8567	2711 8573	2713 8579	2715 8585	2717 8591	2718 8597	0 1	0 1	1 2	1 2	1 3	1 3	1 4	1 5	2 5
74	2720 8602	2722 8608	2724 8614	2726 8620	2728 8626	2729 8631	2731 8637	2733 8643	2735 8649	2737 8654	0 1	0 1	1 2	1 2	1 3	1 3	1 4	1 5	2 5
75	2739 8660	2740 8666	2742 8672	2744 8678	2746 8683	2748 8689	2750 8695	2751 8701	2753 8706	2755 8712	0 1	0 1	1 2	1 2	1 3	1 3	1 4	1 5	2 5
76	2757 8718	2759 8724	2760 8729	2762 8735	2764 8741	2766 8746	2768 8752	2769 8758	2771 8764	2773 8769	0 1	0 1	1 2	1 2	1 3	1 3	1 4	1 5	2 5
77	2775 8775	2777 8781	2778 8786	2780 8792	2782 8798	2784 8803	2786 8809	2787 8815	2789 8820	2791 8826	0 1	0 1	1 2	1 2	1 3	1 3	1 4	1 4	2 5
	0	1	2	3	4	5	6	7	8	9	1	2	3	4	5	6	7	8	9

The first significant figure and the position of the decimal point must be determined by inspection.

SQUARE ROOTS

	0	1	2	3	4	5	6	7	8	9	1	2	3	4	5	6	7	8	9
78	2793 8832	2795 8837	2796 8843	2798 8849	2800 8854	2802 8860	2804 8866	2805 8871	2807 8877	2809 8883	0 1	0 1	1 2	1 2	1 3	1 3	1 4	1 4	2 5
79	2811 8888	2812 8894	2814 8899	2816 8905	2818 8911	2820 8916	2821 8922	2823 8927	2825 8933	2827 8939	0 1	0 1	1 2	1 2	1 3	1 3	1 4	1 4	2 5
80	2828 8944	2830 8950	2832 8955	2834 8961	2835 8967	2837 8972	2839 8978	2841 8983	2843 8989	2844 8994	0 1	0 1	1 2	1 2	1 3	1 3	1 4	1 4	2 5
81	2846 9000	2848 9006	2850 9011	2851 9017	2853 9022	2855 9028	2857 9033	2858 9039	2860 9044	2862 9050	0 1	0 1	1 2	1 2	1 3	1 3	1 4	1 4	2 5
82	2864 9055	2865 9061	2867 9066	2869 9072	2871 9077	2872 9083	2874 9088	2876 9094	2877 9099	2879 9105	0 1	0 1	1 2	1 2	1 3	1 3	1 4	1 4	2 5
83	2881 9110	2883 9116	2884 9121	2886 9127	2888 9132	2890 9138	2891 9143	2893 9149	2895 9154	2897 9160	0 1	0 1	1 2	1 2	1 3	1 3	1 4	1 4	2 5
84	2898 9165	2900 9171	2902 9176	2903 9182	2905 9187	2907 9192	2909 9198	2910 9203	2912 9209	2914 9214	0 1	0 1	1 2	1 2	1 3	1 3	1 4	1 4	2 5
85	2915 9220	2917 9225	2919 9230	2921 9236	2922 9241	2924 9247	2926 9252	2927 9257	2929 9263	2931 9268	0 1	0 1	1 2	1 2	1 3	1 3	1 4	1 4	2 5
86	2933 9274	2934 9279	2936 9284	2938 9290	2939 9295	2941 9301	2943 9306	2944 9311	2946 9317	2948 9322	0 1	0 1	1 2	1 2	1 3	1 3	1 4	1 4	2 5
87	2950 9327	2951 9333	2953 9338	2955 9343	2956 9349	2958 9354	2960 9359	2961 9365	2963 9370	2965 9375	0 1	0 1	1 2	1 2	1 3	1 3	1 4	1 4	2 5
88	2966 9381	2968 9386	2970 9391	2972 9397	2973 9402	2975 9407	2977 9413	2978 9418	2980 9423	2982 9429	0 1	0 1	1 2	1 2	1 3	1 3	1 4	1 4	2 5
89	2983 9434	2985 9439	2987 9445	2988 9450	2990 9455	2992 9460	2993 9466	2995 9471	2997 9476	2998 9482	0 1	0 1	1 2	1 2	1 3	1 3	1 4	1 4	2 5
90	3000 9487	3002 9492	3003 9497	3005 9503	3007 9508	3008 9513	3010 9518	3012 9524	3013 9529	3015 9534	0 1	0 1	0 2	1 2	1 3	1 3	1 4	1 4	1 5
91	3017 9539	3018 9545	3020 9550	3022 9555	3023 9560	3025 9566	3027 9571	3028 9576	3030 9581	3032 9586	0 1	0 1	0 2	1 2	1 3	1 3	1 4	1 4	1 5
92	3033 9592	3035 9597	3036 9602	3038 9607	3040 9612	3041 9618	3043 9623	3045 9628	3046 9633	3048 9638	0 1	0 1	0 2	1 2	1 3	1 3	1 4	1 4	1 5
93	3050 9644	3051 9649	3053 9654	3055 9659	3056 9664	3058 9670	3059 9675	3061 9680	3063 9685	3064 9690	0 1	0 1	0 2	1 2	1 3	1 3	1 4	1 4	1 5
94	3066 9695	3068 9701	3069 9706	3071 9711	3072 9716	3074 9721	3076 9726	3077 9731	3079 9737	3081 9742	0 1	0 1	0 2	1 2	1 3	1 3	1 4	1 4	1 5
95	3082 9747	3084 9752	3085 9757	3087 9762	3089 9767	3090 9772	3092 9778	3094 9783	3095 9788	3097 9793	0 1	0 1	0 2	1 2	1 3	1 3	1 4	1 4	1 5
96	3098 9798	3100 9803	3102 9808	3103 9813	3105 9818	3106 9823	3108 9829	3110 9834	3111 9839	3113 9844	0 1	0 1	0 2	1 2	1 3	1 3	1 4	1 4	1 5
97	3114 9849	3116 9854	3118 9859	3119 9864	3121 9869	3122 9874	3124 9879	3126 9884	3127 9889	3129 9894	0 1	0 1	0 2	1 2	1 3	1 3	1 4	1 4	1 5
98	3130 9899	3132 9905	3134 9910	3135 9915	3137 9920	3138 9925	3140 9930	3142 9935	3143 9940	3145 9945	0 0	0 1	0 1	1 2	1 2	1 3	1 3	1 4	1 4
99	3146 9950	3148 9955	3150 9960	3151 9965	3153 9970	3154 9975	3156 9980	3158 9985	3159 9990	3161 9995	0 0	0 1	0 1	1 2	1 2	1 3	1 3	1 4	1 4
	0	1	2	3	4	5	6	7	8	9	1	2	3	4	5	6	7	8	9

The first significant figure and the position of the decimal point must be determined by inspection.

NAPIERIAN LOGARITHMS $\log_e N$

N	0	1	2	3	4	5	6	7	8	9	1	2	3	4	5	6	7	8	9
1·0	0·0000	0100	0198	0296	0392	0488	0583	0677	0770	0862	10	19	29	38	48	57	67	76	86
1·1	0·0953	1044	1133	1222	1310	1398	1484	1570	1655	1740	9	17	26	35	44	52	61	70	78
1·2	0·1823	1906	1989	2070	2151	2231	2311	2390	2469	2546	8	16	24	32	40	48	56	64	72
1·3	0·2624	2700	2776	2852	2927	3001	3075	3148	3221	3293	7	15	22	30	37	44	52	59	67
1·4	0·3365	3436	3507	3577	3646	3716	3784	3853	3920	3988	7	14	21	28	35	41	48	55	62
1·5	0·4055	4121	4187	4253	4318	4383	4447	4511	4574	4637	6	13	19	26	32	39	45	52	58
1·6	0·4700	4762	4824	4886	4947	5008	5068	5128	5188	5247	6	12	18	24	30	36	42	48	55
1·7	0·5306	5365	5423	5481	5539	5596	5653	5710	5766	5822	6	11	17	23	29	34	40	46	51
1·8	0·5878	5933	5988	6043	6098	6152	6206	6259	6313	6366	5	11	16	22	27	32	38	43	49
1·9	0·6419	6471	6523	6575	6627	6678	6729	6780	6831	6881	5	10	15	20	26	31	36	41	46
2·0	0·6931	6981	7031	7080	7129	7178	7227	7275	7324	7372	5	10	15	20	24	29	34	39	44
2·1	0·7419	7467	7514	7561	7608	7655	7701	7747	7793	7839	5	9	14	19	23	28	33	37	42
2·2	0·7885	7930	7975	8020	8065	8109	8154	8198	8242	8286	4	9	13	18	22	27	31	36	40
2·3	0·8329	8372	8416	8459	8502	8544	8587	8629	8671	8713	4	9	13	17	21	26	30	34	38
2·4	0·8755	8796	8838	8879	8920	8961	9002	9042	9083	9123	4	8	12	16	20	24	29	33	37
2·5	0·9163	9203	9243	9282	9322	9361	9400	9439	9478	9517	4	8	12	16	20	24	27	31	35
2·6	0·9555	9594	9632	9670	9708	9746	9783	9821	9858	9895	4	8	11	15	19	23	26	30	34
2·7	0·9933	9969	**0006**	**0043**	**0080**	**0116**	**0152**	**0188**	**0225**	**0260**	4	7	11	15	18	22	25	29	33
2·8	1·0296	0332	0367	0403	0438	0473	0508	0543	0578	0613	4	7	11	14	18	21	25	28	32
2·9	1·0647	0682	0716	0750	0784	0818	0852	0886	0919	0953	3	7	10	14	17	20	24	27	31
3·0	1·0986	1019	1053	1086	1119	1151	1184	1217	1249	1282	3	7	10	13	16	20	23	26	30
3·1	1·1314	1346	1378	1410	1442	1474	1506	1537	1569	1600	3	6	10	13	16	19	22	25	29
3·2	1·1632	1663	1694	1725	1756	1787	1817	1848	1878	1909	3	6	9	12	15	18	21	25	28
3·3	1·1939	1969	2000	2030	2060	2090	2119	2149	2179	2208	3	6	9	12	15	18	21	24	27
3·4	1·2238	2267	2296	2326	2355	2384	2413	2442	2470	2499	3	6	9	12	15	17	20	23	26
3·5	1·2528	2556	2585	2613	2641	2669	2698	2726	2754	2782	3	6	8	11	14	17	20	22	25
3·6	1·2809	2837	2865	2892	2920	2947	2975	3002	3029	3056	3	5	8	11	14	16	19	22	25
3·7	1·3083	3110	3137	3164	3191	3218	3244	3271	3297	3324	3	5	8	11	13	16	19	21	24
3·8	1·3350	3376	3403	3429	3455	3481	3507	3533	3558	3584	3	5	8	10	13	16	18	21	23
3·9	1·3610	3635	3661	3686	3712	3737	3762	3788	3813	3838	3	5	8	10	13	15	18	20	23
4·0	1·3863	3888	3913	3938	3962	3987	4012	4036	4061	4085	2	5	7	10	12	15	17	20	22
4·1	1·4110	4134	4159	4183	4207	4231	4255	4279	4303	4327	2	5	7	10	12	14	17	19	22
4·2	1·4351	4375	4398	4422	4446	4469	4493	4516	4540	4563	2	5	7	9	12	14	16	19	21
4·3	1·4586	4609	4633	4656	4679	4702	4725	4748	4770	4793	2	5	7	9	12	14	16	18	21
4·4	1·4816	4839	4861	4884	4907	4929	4951	4974	4996	5019	2	5	7	9	11	14	16	18	20
4·5	1·5041	5063	5085	5107	5129	5151	5173	5195	5217	5239	2	4	7	9	11	13	15	18	20
4·6	1·5261	5282	5304	5326	5347	5369	5390	5412	5433	5454	2	4	6	9	11	13	15	17	19
4·7	1·5476	5497	5518	5539	5560	5581	5602	5623	5644	5665	2	4	6	8	11	13	15	17	19
4·8	1·5686	5707	5728	5748	5769	5790	5810	5831	5851	5872	2	4	6	8	10	12	14	16	19
4·9	1·5892	5913	5933	5953	5974	5994	6014	6034	6054	6074	2	4	6	8	10	12	14	16	18
5·0	1·6094	6114	6134	6154	6174	6194	6214	6233	6253	6273	2	4	6	8	10	12	14	16	18
5·1	1·6292	6312	6332	6351	6371	6390	6409	6429	6448	6467	2	4	6	8	10	12	14	16	18
5·2	1·6487	6506	6525	6544	6563	6582	6601	6620	6639	6658	2	4	6	8	10	11	13	15	17
5·3	1·6677	6696	6715	6734	6752	6771	6790	6808	6827	6845	2	4	6	7	9	11	13	15	17
5·4	1·6864	6882	6901	6919	6938	6956	6974	6993	7011	7029	2	4	5	7	9	11	13	15	16

For further values, e.g. $\log_e 4560$, write $4560 = 4{\cdot}560 \times 10^3$ and use the table below.

$\log_e 10^x$

x	1	2	3	4	5	6
$\log_e 10^x$	2·3026	4·6052	6·9078	9·2103	11·5129	13·8155

Printed by Kenya Litho Ltd., Cardiff Road, P.O. Box 775, Nairobi
and published by Oxford University Press, Eastern Africa Branch, National and
Grindlays Bank Building, Government Road, P.O. Box 12532, Nairobi, Kenya